Ethnobotany Investigation of Rice Weeds

Ethnobotany Investigation of Rice Weeds

Shashi Shekhar Prasad

RANDOM PUBLICATIONS
NEW DELHI (INDIA)

Ethnobotany Investigation of Rice Weeds

ISBN 978-93-5111-778-0

Published in 2016 in India by

RANDOM PUBLICATIONS

4376-A/4B, Gali Murari Lal, Ansari Road
New Delhi-110 002
Phone : +9111-43580356, 011-23289044, 011-43142548
e-mail: sales@randompublications.com,
info@randompublications.com, randomexports@gmail.com

Reprinted 2018

Type Setting by : Friends Media, Delhi-110089
Digitally Printed at : Replika Press Pvt. Ltd.

Preface

Rice is the seed of the grass species Oryza sativa (Asian rice) or Oryza glaberrima (African rice). As a cereal grain, it is the most widely consumed staple food for a large part of the world's human population, especially in Asia.

The major impediment in the cultivation of rice is the heavy weed infestation, which compete with the crop to such an extent that the crop gets smoothered by the weeds. The weeds shared not only the plant nutrients but transpire a lot of valuable conserved water from the soil. The weeds also sometime serve as host for breeding and development of certain disease and pests. The delay in first weeding beyond 15-25 days after seeding sharply reduces the rice yield. Rice field weed species may conveniently be classified into broad leaved, grasses, sedges and others.

Management of weeds is an important component of production techniques as elimination of weeds is expansive and hard to achieve. Presence of weeds is a constraint & their improper management further accentuates the effect. In the past two decades work has been done on non-chemical management techniques and environmentally safe alternatives to herbicides for weed control. Such ecofriendly techniques for weed control in rice fields.

Ethnobotanical knowledge encompasses both wild and domesticated species, and is rooted in observation, relationship, needs, and traditional ways of knowing. Such knowledge evolves over time, and is therefore always changing and adding new discoveries, ingenuity and methods.

It is an indispensable book for every researcher, teacher, planner, and students to strengthen the agriculture and food security. Based on his own work over the number of years and that of others before him, the author presents in this book ethnobiological value of rice weeds.

– Author

Contents

Preface .... *v-vi*

1. Introduction to Ethnobotany **1**

Introduction 1
The Field of the Ethnobotanist 4
Ethnobotanical Drug Development 4
Ethnobiological Study of Medicinal Plants 5
Ethnobotanical Importance of Some Plants 7
Definition and Scope f Ethnobotany 12

2. Introduction to Rice **18**

Rice 18
Meiosis and DNA repair 47
Golden Rice 47
Puffed Rice 52
Rice Bran Oil 55

3. Research and Risk-assessment Techniques for Improved Weed Management **59**

Introduction 59
Parameters for Weed-crop Competition 77
Guidelines for Weed-risk Assessment in Developing Countries 94
Resources Required for Weed-Risk Assessment 115

4. Ethnobotanical Investigation of Weed Biodiversity and Rice Production **117**

Background 117
Methods 123
Results and Discussion 125
Ethnobotany and Economic Botany of the North American Flora 131
Weed Biodiversity and Rice Production during the Irrigation Rehabilitation Process 137
Results 141
Discussion 145

5. **Weed Management in Organic Rice** 147

Intercropping 149
Mechanical Methods 150
Mulching & Mowing 151

6. **Utilization of some Weeds as Medicine** 155

Weeds in Agro-ecosystems: As a Source of Medicines for Human Health Care 162

7. **Indigenous Technical Practices in a Rice-based Farming Systems** 168

Indigenous Cropping Systems 168
Indigenous Crop Nutrient Management Practices 170
Indigenous Rice Weed Management Strategies 171
Differential Suppression of Rice Weeds by Allelopathic Plant Aqueous Extracts 172
Seed Wintering and Deteriortion Characteristics beween Weedy and Cultivated Rice 178
Methods 183

8. **Sequence Polymorphisms in Wild, Weedy, and Cultivated Rice** 187

Introduction 187
Materials and Methods 190
Results 191
Discussion 193
Genomic Patterns of Nucleotide Diversity in Divergent Populations of U.S. Weedy Rice 196
Methods 198
The biology of Oryza sativa (Rice) 210

Bibliography 252

Index 254

1

Introduction to Ethnobotany

INTRODUCTION

In aspect of botany which has received recent attention and recognition as an organized discipline in India is ethnobotany, defined as the total direct relationship between humans and plant kingdom. Detailed surveys have been carried out to record the knowledge systems used by tribals and other ethnic communities in the use of plants and other products. Publications such as Contributions to Indian Ethnobotany, Dictionary of Indian Folk Medicine, Ethnobotany and Notable Plants in Ethnomedicine of India provide enormously rich information.

Botany is the study of plants—from the tiniest fern or blade of grass to the tallest or oldest tree. Botany includes all the wild plants and the domesticated species. Domesticates are species that we humans have selected over time from the wild plant species, then tamed and trained to optimally produce for us: food, fibres, medicine, materials, and more. The domesticated species are both the subject and object of agriculture.

Ethnobotanical knowledge encompasses both wild and domesticated species, and is rooted in observation, relationship, needs, and traditional ways of knowing. Such knowledge evolves over time, and is therefore always changing and adding new discoveries, ingenuity and methods. The impacts of modern human societies on traditional cultures and natural habitats have caused huge losses of individual species, and profoundly disrupted communities of species (plant, animal and fungal). Displaced or dispersed peoples—who may have passed along hundreds of generations of observations and customs via oral tradition—lose their languages, the names of things, and their place in the web of relationships. Sometimes new relationships develop as people migrate, and this generates new or modified ethnobotanical knowledge.

WHAT IS ETHNOBOTANY?

Ethnobotany is the study of how people of a particular culture and region make of use of indigenous plants. Ethnobotanists explore how plants are used for such things as food, shelter, medicine, clothing, hunting, and religious ceremonies.

Ethnobotany has its roots in botany, the study of plants. Botany, in turn, originated in part from an interest in finding plants to help fight illness. In fact, medicine and botany have always had close ties. Many of today's drugs have been derived from plant sources.

Pharmacognosy is the study of medicinal and toxic products from natural plant sources. At one time, pharmacologists researching drugs were required to understand the natural plant world, and physicians were schooled in plant-derived remedies. However, as modern medicine and drug research advanced, chemically-synthesized drugs replaced plants as the source of most medicinal agents in industrialized countries. Although research in plant sources continued and plants were still used as the basis for some drug development, the dominant interest (and resulting research funding) shifted to the laboratory.

The 1990's has seen a growing shift in interest once more; plants are reemerging as a significant source of new pharmaceuticals. Industries are now interested in exploring parts of the world where plant medicine remains the predominant form of dealing with illness. South America, for example, has an extraordinary diversity of plant species and has been regarded as a treasure grove of medicinal plants.

The jungles and rain forests of South America contain an incredibly diverse number of plant species, many still unexplored, many unique and potentially useful as medicinal sources. Scientists have also realized the study of the native cultures which inhabit these regions can provide enormously valuable clues in the search for improved health. To uncover the secrets of the rain forest, specialists are needed, well-trained and willing to spend long, hard time in the field. This is where the ethnobotanist comes in.

To discover the practical potential of native plants, an ethnobotanist must be knowledgeable not only in the study of plants themselves, but must understand and be sensitive to the dynamics of how cultures work. Ethnobotanists have helped us to understand the frightening implications which loss of the rain forests would bring not only in terms of consequent loss of knowledge about tropical plants, but the consequent damage brought on by the loss of native cultures in their entirety, as well as the damage to the earth's ecological health.

By necessity, ethnobotany is multidisciplinary. This multidisciplinary approach gives ethnobotanists more insight into the management of tropical forest reserves in a period of tremendous environmental stress. Unfortunately, due to human factors which have influenced the ecological balance of these delicate ecosystems, we are presently faced with the possibility of losing our rain forests.

HOW ARE ETHNOBOTANISTS TRAINED?

Ethnobotanists are usually botanists and/or biologists with additional graduate training in such areas as: archeology, chemistry, ecology, anthropology, linguistics, history, pharmacology, sociology, religion and mythology. With such broad training, ethnobotanists raise many interesting questions quite different in scope from those of previous generations of scientists trained in botany alone. For example, botanists with anthropological and ecological training look at plants as an integral part of human culture. Not only do they study the plants within the tropical forests, they also work respectfully with shamans within the native culture, examining that culture's concepts of disease. With the increased interest in the study of how native peoples use plants therapeutically, pharmacognosy has reemerged as a field of study. Ethnopharmacologists are specifically looking at the pharmacological actions of plants with emphasis on those plants used in treatment of illness, With the renewed interest in using ancient plants as medicinal agents in as well as in religious or sacred activities. Scientists in this field would have a chemistry degree with graduate training in pharmacology and botany.

With the renewed interest in using ancient plants as medicinal agents in modern western medicine, the field of ethnomedicine has emerged. Here physicians receive some cross training in anthropology, botany, public health, or relevant social sciences. These physicians must possess a genuine receptivity to the distinctly unique views of the healing systems practiced by indigenous peoples, as well as the ability to work as a team with ethnobotanists and others.

The physician works with shamans or traditional healers to identify the specific diseases common to both Western cultures and indigenous peoples. In turn, the ethnobotanist works with the shaman to identify and collect plants utilized to treat these diseases.

Following the work of ethnobotanists and physicians trained in ethnomedicine from the field through to research and development of products in pharmaceutical companies can provide us with a glimpse of ethnobotany as it functions today. Let's start with the work that take place before the research expedition begins.

PREPARATION FOR THE EXPEDITION

The first step is collecting detailed knowledge about the local and indigenous people. Researchers prepare a regional study on the epidemiology, traditional medicine, culture and ecology of the people and their environment. In order to prioritize plant collections, a number of international databases are searched to obtain all of the relevant ethnomedical, biological, and chemical information on the plants known to be used in that region. Data is also gathered from remote area hospitals and treatment programs working with local and native peoples. This information is synthesized and integrated into the field research program, the next step of the process.

THE FIELD OF THE ETHNOBOTANIST

Before leaving for field work, ethnobotanists spend many months preparing. They painstakingly gather together the tools and supplies necessary for long-term survival and study in what are often remote villages located deep in dense tropical forests. Here they know they will spend hundreds or thousands of hours in patient observation and experimentation. Ethnobotanists slowly, meticulously, learn about plants the indigenous people use.

They spend long hours cataloging their knowledge about the useful plants and poisonous ones, selecting and collecting plants for cultivation and protection. Above all, ethnobotanists spend long hours completing the repetitious but critical work of pressing and drying plants, often despite monsoon rains and oppressive heat. The plant collection process involves standard methodology, which includes the preparation of multiple plant voucher specimens, which are deposited in the host country as well as in various United States herbaria.

In the field, ethnobotanists work as a team with an ethnomedicine-trained physician to prepare brief case descriptions of diseases. They then present these descriptions of individual diseases to shamans and the local healers, often including photographs of diseases with readily visible clinical manifestations. The interviewing process is conducted very carefully. The cases are presented without using medical terminology. Terms such as herpes, hepatitis, or parasites are not necessarily understood by the local healers. The focus is on common signs and symptoms that are easily recognized. A translator for the local language is usually necessary to conduct this phase. Once a healer has recognized and described the same or similar disease state, the botanical treatment for that condition is recorded in detail by the ethnobotanist. If several independent and reliable shamans describe a similar treatment for a disease, the plant is collected.

Additional plant collections and observations can also be made when Western trained physicians provide health care to the local people. The patient and the local healer are asked about the types of plants used to treat the disease state. Those plants that appear interesting are also collected for later analysis.

ETHNOBOTANICAL DRUG DEVELOPMENT

Once the plants have arrived at the company's research site, processing the plants for medicinal purposes begins. The plants are tagged with the information from the field study. The plants are processed and tested in studies completed by ethnopharmacologists, using state of the art laboratory equipment (which may include High Pressure Liquid Chromatography studies and in vivo transgenic animal studies). The objective is to screen the plants metabolites to determine how relevant they are to the therapeutic areas of interest. The most promising initial plant compounds are fractionated to obtain pure samples in milligram amounts. These natural pure compounds are compared to the best available therapeutics by in vitro testing. If the bioassay is successful, the compound is structurally characterized and is subject to a confirmatory biological test.

Promising compounds are scaled up to provide gram quantities for animal testing to determine safety and efficacy. Pharmacologists with backgrounds in metabolism, pharmacokinetics, medicinal chemistry and formulation design experiments to determine whether the selected compounds have activity. After this testing is completed, the samples are compared with the best available marketed therapeutics. The scale up process occurs again and hundreds of grams of selected compounds are provided for further studies which will, it is hoped, eventually lead to an effective, marketable drug suitable for human consumption. This new product is the reward for all the time and effort of many individuals.

ETHNOBIOLOGICAL STUDY OF MEDICINAL PLANTS

Medicinal plants as a group comprise approximately 8000 species and account for around 50 per cent of all the higher flowering plant species of India. Millions of rural households use medicinal plants in a self-help mode. Over one and a half million practitioners of the Indian System of Medicine in the oral and codified streams use medicinal plants in preventive, promotive and curative applications.

There are estimated to be over 7800 manufacturing units in India. In recent years, the growing demand for herbal product has led to a quantum jump in volume of plant materials traded within and across the countries. An estimate of the EXIM Bank puts the international market of medicinal plants related trade at US$ 60 billion per year growing at the rate of 7 per cent only. Though India has a rich biodiversity, the growing demand is putting a heavy strain on the existing resources.

While the demand for medicinal plants is growing, some of them are increasingly being threatened in their natural habitat. For meeting the future needs cultivation of medicinal plant has to be encouraged.

According to an all India ethnobiological survey carried out by the Ministry of Environment & Forests, Government of India, there are over 8000 species of plants being used by the people of India.

MEDICINAL PLANTS AS A PART OF CULTURE

It is evident that the Indian people have a tremendous passion for medicinal plants and use them for a wide range of health related applications from a common cold to memory improvement and treatment of poisonous snake bites to a cure for muscular distrophy and the enhancement of body's general immunity.

In the oral traditions local communities in every ecosystem from the trans Himalayas down to the coastal plains have discovered the medical uses of thousands of plants found locally in their ecosystem. India has one of the richest plant medical culture in the world. It is a culture that is of tremendous contemporary relevance because it can on one hand ensure health security to millions of people and on the other hand it can provide new and safe herbal drugs to the entire world. There are estimated to be around 25000 effective plant based formulations are available in the indigenous medical texts.formulations used in folk medicine and known to rural communities all over India and around 10000 designed.

Distribution

Macro analysis of the distribution of medicinal plants show that they are distributed across diverse habitats and landscape elements. Around 70 per cent of India's medicinal plants are found in tropical areas mostly in the various forest types spread across the Western and Eastern ghats, the Vindhyas, Chotta Nagpur plateau, Aravalis & Himalayas. Although less than 30 per cent of the medicinal plants are found in the temperate and alpine areas and higher altitudes they include species of high medicinal value. Macro studies show that a larger percentage of the known medicinal plant occur in the dry and moist deciduous vegetation as compared to the evergreen or temperate habitats.

Analysis of habits of medicinal plants indicate that they are distributed across various habitats. One third are trees and an equal portion shrubs and the remaining one third herbs, grasses and climbers. A very small proportion of the medicinal plants are lower plants like lichens, ferns algae, etc. Majority of the medicinal plant are higher flowering plants.

DISTRIBUTION OF MEDICINAL PLANTS BY HABITS

Of the 386 families and 2200 genera in which medicinal plants are recorded, the families Asteraceae. Euphorbiacae. Laminaceae, Fabaceae, Rubiaceae., Poaceae, Acanthaceae, Rosaceae and Apiaceae share the larger proportion of medicinal plant species with the highest number of species (419) falling under Asteraceae.

About 90 per cent of medicinal plant used by the industries are collected from the wild. While over 800 species are used in production by industry, less than 20 species of plants are under commercial cultivation. Over 70 per cent of the plant collections involve destructive harvesting because of the use of parts like roots, bark, wood, stem and the whole plant in case of herbs.

This poses a definite threat to the genetic stocks and to the diversity of medicinal plants if biodiversity is not sustainably used.

ETHNOBOTANICAL IMPORTANCE OF SOME PLANTS

GARLIC (*LAHSAN, ALLIUM SATIVUM*)

Garlic belongs to the Liliaceac's family. Garlic is believed to have originated in Central Asia and Mediterranean. It is a perennial plant with narrow flat leaves and several small egg shaped bulbs, known as cloves, enclosed within the white skin. The characteristic odour is due to allicin, chemically known as diallyl disulphide.

Uses:

(1) The bulblets together with the cloves and the leaves, have been used from earliest time for flavouring soups, sausages, and salads.

(2) Garlic also has a medicinal value as it possesses antiseptic, and bactericidal properties.

(3) Garlic powder is extensively used as condiment and also serves as carminative and gastic stimulant.

(4) Bulbs are also used in treating cough fever and rheumatism. Garlic juice is useful in skin diseases.

(5) The seeds yields an aromatic oil which is very useful for checking cold fits of intermittent fevers, externally this oil is employed as a liniment for paralytic and rheumatic affections.

Garlic is considered useful in case of snake bite.

Aloe (*Ghi Kanvar, Aloe vera*)

This juicy shrub is a native of the West Indies. Several kinds of aloes are on the market. Barbados or curacao aloes came from *Aloe barbadensis* = *A. vera*. The leaves contain a resinous juice in which there are several glucosides. If the leaves are cut and placed in troughs, the juice slowly oozes out and can be collected.

Uses:

(1) Aloe is used chiefly as purgatives.

(2) Medicinally, the plant is used in the treatment of piles and rectal fissures.

(3) Pulp of the leaves is used in the treatment of irregularity in menustrual cycle.

Margosa, Neem *(Azadirachta indica)*

It is a common tree of Meliaceae, throughout the greater part of India and often planted for shade.

Uses:

(1) The plant has great medicinal value.

(2) The seeds are the sources of margosa oil which is used skin diseases, applied as a linament in rheumatism and possesses anthelmintic and insecticidal properties.

(3) It is also used in bathing soaps and tooth paste.

(4) The oil cake is a useful fertilizer.

(5) The bark is very bitter and is beneficial in malarial fever.

(6) The leaves are bitter and posses insecticidal properties. The dry flowers are tonic and stomachic.

(7) The tender twigs are used to clean teeth and are very effective pyorrhea.

Indian Laburnum (*Amaltas, Cassia fistula)*

The plant is a member of sub-family Caesalpiniaceae of family Leguminosae. Cassia is native to arid regions of Egypt and Arabia. Species of Cassia are cultivated in India.

Uses:

(1) An ancient drug 'senna' is obtained from the dried leaflets, and also the pods.

(2) The leaves are picked by the natives, dried in the sun, Senna is used as laxative vermifuge, cathartic and as a purgative.

(3) Bark of the plant is used in tanning of leather.

(4) Timber of the plant is used in making agricultural tools. The plant is known as 'golden shower' for its golden yellow flowers.

Emblic (*Amla, Emblica officinalis*)

Amla or Aonla (Euphorbiaceae), is a moderate sized tree found throughout India.

Uses:

(1) Fruits are picked and also eaten with sugar in the form of 'murabba'.

(2) They are rich suorce of Vitamin 'C' and are a good liver tonic.

(3) Amla is abundantly used in Ayurvedic medicinal system.

(4) Amla is one of the chief components in preparation of two famous Ayurvedic medicines. Trifala and Chayavanprash.

(5) Dried fruits of amla has detergant property, therefore, they are used as shampoo.

(6) Amla fruits are used in preparation of hair dye.

(7) The fruits are also used in the treatment of scurvy disease.

Castor (*Arandi Ricinus communis*)

R. communis seeds are the source of a very important industrial non-drying oil, the castor oil. Castor belongs to the family Euphorbiaceae. The plant is native of Africa but grows in wild as well as cultivated state in the whole world in both the temprate and topical regions. It is chiefly cultivated in Brazil, India, Mexico, U.S.S.R. and Sudan. The plant is an erect herbaceous annual shrub. The seeds contain 35 to 55 per cent oil.

Uses:

(1) The castor oil is of medicinal value.

(2) It acts as a purgative.

(3) The oil has a number of industrial uses, It is used in the manufacture of soaps, inks, plastics, imitation leather, linoleum, nylon *etc.*

(4) The oil is a very good lubricant.

(5) As a lubricant it is used in a number of machineries.

(6) The oil is used in a number of articles because of its water resistant quality.

(7) It is used for preserving leather.

(8) Castor oil becomes a drying oil if it is heated and dehydrated. It can then be mixed with more expensive oils in the paint industry.

(9) The oil is also used as a fuel in lamps *etc.*

(10) The oil cake is poisonous but is a good fertilizer.

(11) The stem is a source of paper pulp and cellulose.

Terminalia *(Terminalia sp.)*

Terminalia is an exclusively tropical plant and about a dozen species occur in India. Important species are— *Terminalia chebula* (Black myrobalans) or Hararh, *Terminalia bellerica* (Belleric myrobalans) or Behera, *Terminala arjuna* (Arjuna myrobalans) or Arjun.

The fruits of these three species figure in the international commerce. They are largely used in Indian system of medicine as they have laxative, stomachic, and tonic properties. The fruit extract of *T. bellerica* also possesses antibacterial properties. An important Ayurvedic drug Triphala is prepared from harah, behera and amla. The nuts of *T. chebula* and *T. bellerica* have a tannin content of 30 to 40 percent. When used alone they yield a spongy leather of a light yellow colour, but in combination they are more satisfactory. They are used in the tanning of calf, goat, and sheep leather. *T. arjuna* and *T. chebula* are also used for dyeing and ink.

Ginger (*Adarakh, Zingiber officinale*)

Ginger belongs to the family Zingiberaceae. It orginated in tropical south-east Asia. China and India were the first countries to grow it. For many years it was an important drug. It was the principal ingredient of a remedy for the plague. Today ginger is cultivated over a wider area than most species. The ginger plant is an erect perennial herb with thick scaly rhizomes that branch digitately. It is cultivated for the most part in small home gardens. The plant is perpetuates by the rhizomes. The rhizomes are pale yellow externally and greenish yellow inside. They contain starch, gums, an oleoresin, and an essential oil as well. The rhizomes are dug after the aerial parts of the plant have withered.

Ginger is prepared in two different ways—

(1) Preserved or green ginger is a product of southern China. Young juicy rhizomes are dried cleaned and boiled in water until soft. They are then peeled, scraped and boiled several times in a sugar solution, and finally packed in a similar solution.

(2) Dried or cured ginger is product of the other ginger-growing countries. The rhizomes are cleaned, peeled, and dried in the sun. They are sometimes parboiled in water or lime juice before peeling. This is black ginger. White ginger is made by bleaching the rhizomes.

The aromatic colour of ginger is due to the presence of essential oil, while the pungent taste is due to the presence of the nonvolatile oleoresin, gingerin. Ginger is used more as a codiment than as a spice. It expands the blood vessels in the skin, causing a feeling of warmth, and increase perspiration, with an accompanying drop in temperature. For this reason it is much used in warm countries.

Holy Basil (*Tulsi, Ocimum sanctum*)

A perennial aromatic herb of the family Labiatae, it is grown in specially made pots or pedestelins in orthodox Hindu homes. There are two varieties one with green leaves (Saritulsi) and the second with purple leaves (Krishnatulsi). The plant is very sacred in India. The terms Haripriya for it means dear to Lord Vishnu. People believe the evil spirits cannot haunt a place where Tulsi is planted. The plant is worshipped in morning and in evening by placing an oil and 'Ghee' lamp before it. The worship of Tulsi reached every corner in India through Vaishnavism. In Bengal, with the preaching of Chaitanya cult, the worship of Tulsi became universal. It is believed that water with few leaves of Tulsi becomes as holy as water of the Ganges. The plant is never burnt by the Hindus. Small pieces of wood are used as beads in rosaries and necklace by devotes of Krishna. This is believed to remove sins.

It is a native of India and Africa.

Uses:

(1) The juice of its leaves is the common cure of colds.

(2) Infusion of leaves is useful in catarrah, bronchitis and digestive complaints.

(3) Oil from the leaves has antibacterial property and is applied on ringworm and other skin diseases.

(4) The plant is mosquito repellent also. A volatile oil is obtained from the leaves of basil. It contain cineole, 1-linalool, methyl cinnamate, eugenol *etc.*

(5) Roots of tulsi are used as febrifuge and as an antidote to snake bite.

Safed Musli (*Chlorophytum borivilianum*)

The plant is the member of family Liliaceae. It is found in W. Himachal, Punjab, Madhya Pradesh and Central India. It is a shrub. Roots of the plant are medicinally important and are used in the treatment of general weakness, spermetorrhoea, impotancy, leucorrhoea, dysentry diarrhoea, rheumatism *etc.* Roots are also used as tonic.

Ashwagandha (*Withania somnifera*, Winter Cherry)

The plant is a member of family Solanaceae and is a small or middle-sized undershrub, upto 1.5m high. It is common in drier parts of India.

Uses:

(1) The leaves and roots are used medicinally.

  (a) The leaves are used as an anthelmintic and febrifuge.

  (b) An infusion is given in fevers.

  (c) for the treatment of piles a decoction of the leaves is used bot internally and externally.

  (d) The leaves are also used as a hypotonic in alcoholism

(2) An infusion of bark is given in asthama.

(3) The root contains the alkaloid somniferine.

  (a) The root powder is applied locally on ulcers and inflammations.

(4) Ashwagandha is useful in sexual and general weakness and rheumatism.

(5) The berries and seeds are dieuretic, they are used in chest complaints.

(6) A paste made of the berries is an effective ointment for ringworm.

(7) The antibiotic and antibacterial activity of the roots as well as leaves has recently been shown experimentally.

Guggal (*Commiphora wightii, Indian Bdellium*)

The plant is the member of family Burseraceae. It is a small thorny tree or shrub distributed in Rajasthan, Mysore, Bengal and Madhya Pradesh.

Uses: The aleo-gum-resin from the tree is of considerable medicinal value. The gum is the Indian bdellium, the guggal of the Indian Materia Medica. The fresh gum is widely used for a variety of diseases.

(a) It is considered a bitter stomactic, carminative, demulcent, astringent, and antiseptic. It is also used as diaphoretic, diuretic and expectorant.
(b) Internally it is prescribed as an uterine stimulant, it regulates the menstrual functions.
(c) In large doses it is given in leucorrhoea.
(d) As an intestinal disinfectant it is prescribed for diseases like chronic catarrh of the bowels, chronic colitis, diarrhoea and tubercular ulceration of the bowels.
(e) As an expectorant it is valuable in pulmonary tuberculosis, in large doses repeated four or five times a day it is recommended for bronchitis, whooping cough and pneumonia.
(f) For the treatment of rheumatism it is considered valuable.
(g) As a mouth wash and gargle the gum is given for weak and spongy gums, chronic tonsilitis, ulcerated throat and pyorrhoea.
(h) Fumes arising from the burning gum are inhaled in hay fever and chronic laryngitis.

DEFINITION AND SCOPE OF ETHNOBOTANY

Ethnobotanical knowledge is very ancientg in India and might be among the earliest in the world and all traditional systems of medicine had their roots in ethnobotany. Yet, organized studies in ethnobotany are very recent. Ethnobotany can be defined as the total natural and traditional relationship and the interactions between man and his surrounding plant wealth. Ethnobotany must have been the first knowledge which the early man had acquired out of necessity, intutition, observation and experimentation.

Archaeological or palaeobotanicla evidences about collection, use and cultivation of any plant products by early man for food, house building *etc.* and references to herbal medicines in ancient scriptures suggest a very long history of ethnobotany. The word 'ETHNOBOTANY' was applied to such knowledge by Harshberger (1895) less than a century ago and till a few years ago the only book on the subject was Introduction to Ethnobotany by Faulks (1958). According to *Schultes* (1962), ethnobotany is "The study of the relationship which exist between people of primitive societies and their plant enviornment." In more simple words, it is an anthropological approach to botany. Majumdar (1938) wrote a book of the title " Some Aspects of Indian Civilization". Systematic field and related studies in the subject pioneered by Botanical Survey of India. E.K. Janaki Ammal studied subsistence food plants of certain tribals of South India. Faulks (1958) almost deals with whole economic botany. Researches of Schultes (1962) in Amazon and Barrau (1959,66) and Conklin (1962) in Melanesia and South- East Asia greatly prompted Indian Workers into Ethnobotany.

Jain (1981) undertook intensive field study among the tribals of Central India, and also devised methodology for ethnobotany, particularly in the Indian context. The publications from his group in the early sixties triggered ethnobotany activity in many other centres, particularly among botanists, anthropologists and Ayurvedic medical practitioners. During the last two decades, work has started at, National Botanical Research Institute (NBRI), Lucknow, National Bureau of Plant Genetic Resources (NBPGR), Delhi, Central Institute of Medicinal and Aromatic Plants (CIMAP), Central Council for Research in Ayurveda and Siddha (CCRAS) and Central Council for Research in Unani Medicine (CCRUM).

Several Botany Departments of Univrsities are also associated with this type of work.

In recent years, much work in this science has been done in U.S.A., England, France, Mexico, India *etc.* An All India Co-ordinated Research Project on Ethnobiology was initiated in 1974. It took some years to finalise the work programme and the project came into operation from 1982, at NBRI Lucknow, four centres of Botanical Survey of India (Shillong, Howrah, Coimbatore and Port Blair) and some other institutions.

According to Faulks (1958), the following informations conerning the plants used by the ethnic or tribal people are collected by an ethnobotanist.

(i) Plants concerning domestic and wild animals—The plants used as veterinary medicines, foods and fodder for the animals and to control the wild animals and pests by using plant poison *etc.*

(ii) Plants used as food vegetable and fruits—Cultivated as well as the wild ones.

(iii) Plants used as fats and oil, flavour and condiments.

(iv) Plants of drinks, beverage, fumitories.

(v) Household equipments and furniture—The plants which provide shelter and protection, furniture, wall carvings, water mills *etc.*

(vi) Plants that provide fuel and manure.

(vii) Plant that provide clothing.

(viii) Plants used as cosmetics.

(ix) Plants used for packaging material and cordage.

(x) Plants associated with recreation and instruments— The Plants used for making articles of play, walking sticks, toys, musical instruments *etc.*

(xi) Trade—The plants extracted from the forests for trade purpose such as medicinal, timber and grass *etc.*

(xii) Plants which are related to medicine, charms and hullucinations *etc.*

Besides these, the primitive ideas, thoughts and concepts, folklore stories or songs, abuses or blessings conerning the plants are also recorded along with native names and phonetic transactions.

To collect such a vast information from the ethnic or tribal regions, required both time and patience.

And, this type of work cannot be fully accomplished without living for long periods in the area of study. Few professional Indian botanists or ethnobotanists can get the facilities and privilage like Dr. Richard Evans Schultes, who spent an almost uninterrupted 12 years in North West Amazon, and published valuable work. The important ethnobotanical works in India have been enumrated by Jain (1979).

Ethnobotanical data are collected from several sources like—

(a) Archaeological Sources—India has a rich treasure of archaelogical sculptues of antiquity, which can be of great help in tracing the plants which were used during early civilization.

(b) Literature as Source—Our ancient literature can be a source of information on important ethnic plants.

(c) Herbarium as Source—Herbarium sheet and field notes have also proved to be a good source of ethnobotanical data. The most outstanding example of this type of research is of Dr. Altschul, who researhed about 2.5 million plant speciments in Harvard Universities Herbarium and from these 5,178 useful notes of drugs and food value were recorded.

(d) Field as Source—Ethnobotanist brings out from the field the suggestion as to which raw plant material may be tapped and for this, he gets clues from the tribals.

SUB-DIVISIONS OF ETHNOBOTANY

Ethnobotanical studies on various sub-groups of the plant kingdom, like a algae, fungi, bryophytes, pteridophytes, lichens *etc.* are linked under the sub heads of ethnoalgology, ethnomycology, ethnobryology, ehnopteridology, ehnolichenology *etc.*

Studies on special aspects of botany like systems of classification, medicinal uses palaeobotany, ecology *etc.* are also subdisciplines and have been termed as ethnotaxonomy, ethnomedicobotany, ethnoecology, palaeoethnobotany, *etc.*

But when the inquiry in ehtnobotany extends beyond ordinary realm of botany and has significant input of another branch of science, like archaeology or medicine, the work becomes interdisciplinary. Now a days ethobotany is an interdisciplinary subject and it has many sub-divisions, which are following—

(a) Ethnotaxonomy—It includes the study of the nomenclature, identification, classification of the ethnobotanically important plants.

Brown (1984) has given a Folk Biological classification which was based on the dialect of local people.

(b) Archaeoethnobotany—It includes the study of human civilization and crop evolution with the help of archaeological materials.

(c) Ethnoecology—It includes the study of mutual relationships of human beings of past and present with their biotic and abiotic environment.

(d) Ethnopharmocology—It includes the identification and description of medicinally important plants. It also deals with the study of effects of the medicines on human beings, which are extracted from these plants.

(e) Ethnomusicology—It includes the study of the understanding of music of tribal people.

(f) Ethnotoxicology—It includes the study of toxic plants and the toxic substances which are obtained from them. These toxic substances are used as fish poison and arrow poison by many tribes (Schultes 1970).

(g) Palaeoethnobotany—It deals with the study of fossils of ancient plants, specially plants of economic importance. The fossils of crops give the idea about the culture, level of farming, *etc.* of ancient human beings.

(h) Ethnogynacolgy—It is a new discipline of ethnobotany. It includes the study of the diseases in tribal women and their treatment with the help of plants.

(i) Ethnonarcotics—This branch includes the study of intoxicants which are used by the tribals and ancient people, for example-snuff, opium, hellucinogens *etc.*

(j) Ethnomycology—It includes the study of fungal diseases and their treatment, preparation of beverages with the help of fungi and food products obtained from fungi by ancient and tribal people.

SCOPE OF ETHNOBOTANY

There are many lines of study or work in ethnobotany, some important of them are as follows—

(1) Much ethnobotanical work is involved to study and describe articles of domestic or professional use, including the huts or houses of the adivasis. Researchers have found much ingenuity in design, choice of timber or other materials and in art of making and decorating these articles.

(2) One interesting area of work in ethnobotany is the study of selection brought about by the tribals in certain economic plants, and the conservation of germ plasm through patronage of land races. Many

tribals and other orthodox cultivators have not adopted all new or improed varieties of crops and have continued to grow the traditional (even less productive) commonly cultivated crops, there by maintaining their genetic material or germplasm. Special characters of hardiness, disease resistance and adaptability to special conditions like water logging, extreme drought or cold, *etc.* in the land races have been continuously utilized by the plant breeders and on some occasions helped averting widespread famines. Arora and his associates (1981,84,87) have done tremendous work in India and shown infinite properties of such work on not only food plants, but all groups of economic plants.

(3) Ethnobotanical studies help in revealing the numerous germplasm stocks of our cultivated plants and vegetables. The tribals living in interiors of forests also cultivate numerous vegetable crops for generations and thus represent a distinct genetic stock adapted to local conditions, for example *Piper peepuloides* used as condiment in Khasi hills, *Moghania vestita* of Khasi and Jaintia hills, as tuber crop, and others.

(4) The study of origin or basis of local or vernacular names of plants is also a part of ethnobotanical study. Useful plants, like any other objects, need to be referred frequently and are assigned names. Vernacular or local names vary with place, language and people, they are useful, simple and easy for the local people to pronounce and queite often have relevant meaning.

(5) The sacred forest and trees of India have not been explored by active field botanists. Closer analysis is necessary of the ecological impacts of these traditions in different locations. All kinds of mythological associations with plants need to be recorded and analysed.

(6) Concern about the increasing population, particularly in many developing countries, directed research into traditional methods of contraception and ethnobotanical data. Many plants were found to be having such qualities. Kamboj (1982) reviewed the work on plants for fertility regulation in India.

(7) Many naturally growing or wild plants still supply food to large sections of human beings. In remote mountains or desertic region and among unculturated human societies, wild plant products (tuber, leaves, flowers, fruits, seeds, grains, *etc.*) provide considerable quantities of food.

Ethnobotany is a very vast subject. The reasons for this more and more wide horizons of ethnobotany are primarily three:

(1) The origin and soul of this subject are in primary basic human needs in tradition and in faith, things, which are immortal and infinite.

(2) Secondly, the body and substance of this science are in plant resources—plants which sustain all life on this earth, and are a renewable resource.

(3) Thirdly, and most significantly, the subject has close historical implications with our past, relation with so many facets of our present thought, responses, needs and actions and relevance also with our future welfare and environments- like needs of new foods and medicines and conservation of genetic diversity or ecosystems.

2

Introduction to Rice

RICE

Rice is the seed of the grass species Oryza sativa (Asian rice) or Oryza glaberrima (African rice). As a cereal grain, it is the most widely consumed staple food for a large part of the world's human population, especially in Asia. It is the agricultural commodity with the third-highest worldwide production, after sugarcane and maize, according to 2012 FAOSTAT data.

Fig. A mixture of brown, white, and red indica rice, also containing wild rice, Zizania species

Fig. Oryza sativa with small wind pollinated flowers

Since a large portion of maize crops are grown for purposes other than human consumption, rice is the most important grain with regard to human nutrition and caloric intake, providing more than one fifth of the calories consumed worldwide by humans.

Fig. Cooked brown rice from Bhutan

Chinese legends attribute the domestication of rice to Shennong, the legendary Emperor of China and inventor of Chinese agriculture. Genetic evidence has shown that rice originates from a single domestication 8,200–13,500 years ago in thePearl River valley region of China. Previously, archaeological evidence had suggested that rice was domesticated in theYangtze River valley region in China.

From East Asia, rice was spread to Southeast and South Asia. Rice was introduced to Europe through Western Asia, and to the Americas through European colonization.

Fig. Rice can come in many shapes, colours and sizes. Photo by the IRRI.

There are many varieties of rice and culinary preferences tend to vary regionally. In some areas such as the Far East or Spain, there is a preference for softer and stickier varieties.

Rice, a monocot, is normally grown as an annual plant, although in tropical areas it can survive as a perennial and can produce a ratooncrop for up to 30 years. The rice plant can grow to 1–1.8 m (3.3–5.9 ft) tall, occasionally more depending on the variety and soil fertility. It has long, slender leaves 50–100 cm (20–39 in) long and 2–2.5 cm (0.79–0.98 in) broad. The small wind-pollinated flowers are produced in a branched arching to pendulous inflorescence 30–50 cm (12–20 in) long. The edible seed is a grain (caryopsis) 5–12 mm (0.20–0.47 in) long and 2–3 mm (0.079–0.118 in) thick.

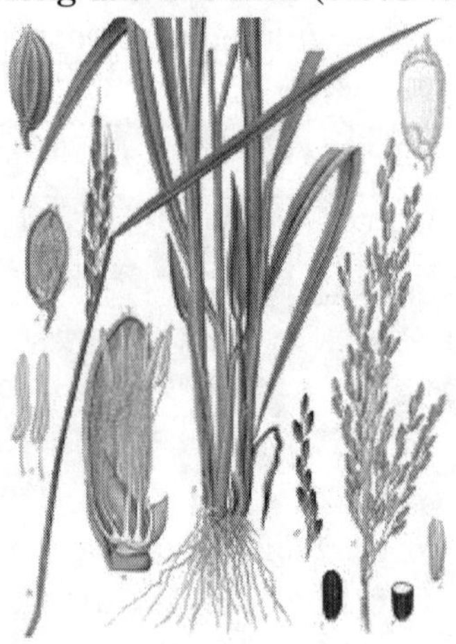

Fig. Oryza sativa, commonly known as Asian rice

Rice cultivation is well-suited to countries and regions with low labor costs and high rainfall, as it is labor-intensive to cultivate and requires ample water. However, rice can be grown practically anywhere, even on a steep hill or mountain area with the use of water-controlling terrace systems. Although its parent species are native to Asia and certain parts of Africa, centuries of trade and exportation have made it commonplace in many cultures worldwide.

The traditional method for cultivating rice is flooding the fields while, or after, setting the young seedlings. This simple method requires sound planning and servicing of the water damming and channeling, but reduces the growth of less robust weed and pest plants that have no submerged growth state, and deters vermin. While flooding is not mandatory for the cultivation of rice, all other methods ofirrigation require higher effort in weed and pest control during growth periods and a different approach for fertilizing the soil.

The name wild rice is usually used for species of the genera Zizania and Porteresia, both wild and domesticated, although the term may also be used for primitive or uncultivated varieties of Oryza.

ETYMOLOGY

First used in English in the middle of the 13th century, the word "rice" derives from the Old French ris, which comes from Italian riso, in turn from the Latin oriza, which derives from the Greek ????a (oruza). The Greek word is the source of all European words (cf. Welsh reis, German Reis, Lithuanian ry•iai, Serbo-Croatian ri•a, Polish ryz, Dutch rijst, Hungarian rizs, Romanian orez).

The origin of the Greek word is unclear. It is sometimes held to be from the Tamil word (arisi), or rather Old Tamil arici. However, Krishnamurti disagrees with the notion that Old Tamil arici is the source of the Greek term, and proposes that it was borrowed from descendants of Proto-Dravidian *wariñci instead. Mayrhofer suggests that the immediate source of the Greek word is to be sought in Old Iranian words of the types *vriz- or *vrinj-, but these are ultimately traced back to Indo-Aryan (as in Sanskrit vrihí-) and subsequently to Dravidian by Witzel and others.

COOKING

The varieties of rice are typically classified as long-, medium-, and short-grained. The grains of long-grain rice (high in amylose) tend to remain intact after cooking; medium-grain rice (high in amylopectin) becomes more sticky. Medium-grain rice is used for sweet dishes, for risotto in Italy, and many rice dishes, such as arròs negre, in Spain. Some varieties of long-grain rice that are high in amylopectin, known as Thai Sticky rice, are usually steamed. A stickier medium-grain rice is used for sushi; the stickiness allows rice to hold its shape when molded. Short-grain rice is often used for rice pudding.

Instant rice differs from parboiled rice in that it is fully cooked and then dried, though there is a significant degradation in taste and texture. Rice flour and starch often are used inbatters and breadings to increase crispiness.

Preparation

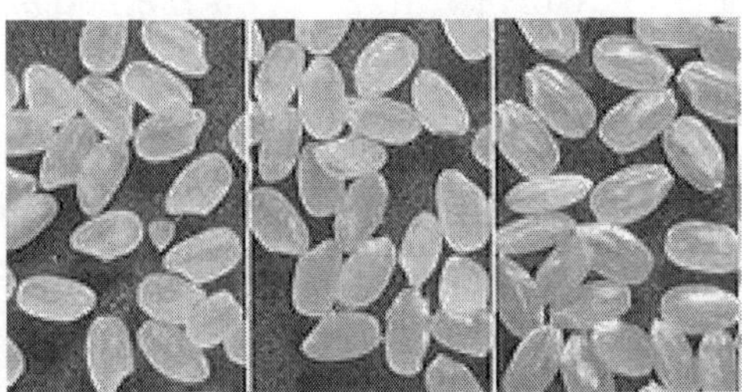

Fig. Milled to unmilled rice, from left to right, white rice (Japanese rice), rice with germ, brown rice

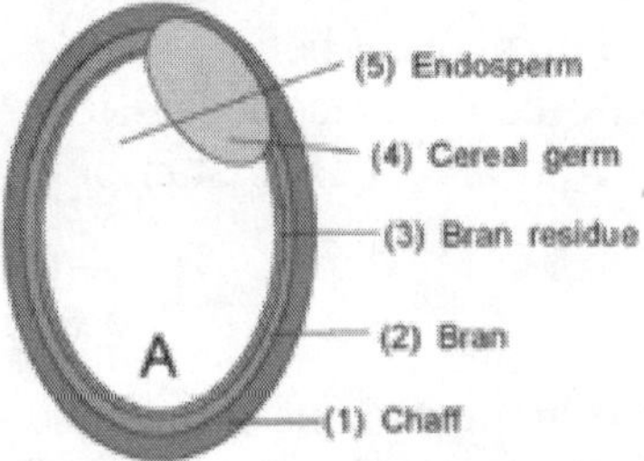

- Rice with chaff

- Brown rice
- Rice with germ
- White rice with bran residue
- Musenmai (Japanese: ???), "Polished and ready to boil rice", literally, non-wash rice
- Chaff
- Bran
- Bran residue
- Cereal germ
- Endosperm

Rice is typically rinsed before cooking to remove excess starch. Rice produced in the US is usually fortified with vitamins and minerals, and rinsing will result in a loss of nutrients. Rice may be rinsed repeatedly until the rinse water is clear to improve the texture and taste.

Rice may be soaked to decrease cooking time, conserve fuel, minimize exposure to high temperature, and reduce stickiness. For somevarieties, soaking improves the texture of the cooked rice by increasing expansion of the grains. Rice may be soaked for 30 minutes up to several hours.

Brown rice may be soaked in warm water for 20 hours to stimulate germination. This process, called germinated brown rice (GBR),activates enzymes and enhances amino acids including gamma-aminobutyric acid to improve the nutritional value of brown rice. This method is a result of research carried out for the United Nations International Year of Rice.

Processing

Rice is cooked by boiling or steaming, and absorbs water during cooking. With the absorption method, rice may be cooked in a volume of water similar to the volume of rice. With the rapid-boil method, rice may be cooked in a large quantity of water which is drained before serving. Rapid-boil preparation is not desirable with enriched rice, as much of the enrichment additives are lost when the water is discarded. Electric rice cookers, popular in Asia and Latin America, simplify the process of cooking rice. Rice (or any other grain) is sometimes quickly fried in oil or fat before boiling (for example saffron rice or risotto); this makes the cooked rice less sticky, and is a cooking style commonly called pilaf in Iran and Afghanistan or biryani (Dam-pukhtak) in India and Pakistan.

Dishes

In Arab cuisine, rice is an ingredient of many soups and dishes with fish, poultry, and other types of meat. It is also used to stuff vegetables or is wrapped in grape leaves (dolma). When combined with milk, sugar, and honey, it is used to make desserts. In some regions, such as Tabaristan, bread is made using rice flour. Medieval Islamic texts spoke of medical uses for the plant. Rice may also be made into congee (also called rice porridge, fawrclaab, okayu, Xifan, jook, or rice gruel) by adding more water than usual, so that the cooked rice is saturated with water, usually to the point that it disintegrates. Rice porridge is commonly eaten as a breakfast food, and is also a traditional food for the sick.

NUTRITION AND HEALTH

Nutrients and the nutritional importance of rice

Rice is the staple food of over half the world's population. It is the predominant dietary energy source for 17 countries in Asia and the Pacific, 9 countries in North and South America and 8 countries in Africa. Rice provides 20% of the world's dietary energy supply, while wheat supplies 19% and maize (corn) 5%.

A detailed analysis of nutrient content of rice suggests that the nutrition value of rice varies based on a number of factors. It depends on the strain of rice, that is between white, brown, red, and black (or purple) varieties of rice – each prevalent in different parts of the world. It also depends on nutrient quality of the soil rice is grown in, whether and how the rice is polished or processed, the manner it is enriched, and how it is prepared before consumption.

An illustrative comparison between white and brown rice of protein quality, mineral and vitamin quality, carbohydrate and fat quality suggests that neither is a complete nutrition source. Between the two, there is a significant difference in fiber content and minor differences in other nutrients.

Highly colored rice strains, such as black (purple) rice, derive their color from anthocyanins and tocols. Scientific studies suggest that these color pigments have antioxidant properties that may be useful to human health. In purple rice bran, hydrophilic antioxidants are in greater quantity and have higher free radical scavenging activity than lipophilic antioxidants. Anthocyanins and ?-tocols in purple rice are largely located in the inner portion of purple rice bran.

Comparative nutrition studies on red, black and white varieties of rice suggest that pigments in red and black rice varieties may offer nutritional benefits.

Red or black rice consumption was found to reduce or retard the progression of atherosclerotic plaque development, induced by dietary cholesterol, in mammals. White rice consumption offered no similar benefits, which the study suggests may be due in part to a lack of antioxidants found in red and black varieties of rice.

Comparison of rice to other major staple foods

Raw grains, however, are not edible and can not be digested. These must be sprouted, or prepared and cooked for human consumption. In sprouted and cooked form, the relative nutritional and anti-nutritional contents of each of these grains is remarkably different from that of raw form of these grains reported in this table.

[A] corn, yellow	[B] rice, white, long-grain, regular, raw, unenriched
[C] wheat, hard red winter	[D] potato, flesh and skin, raw
[E] cassava, raw	[F] soybeans, green, raw
[G] sweet potato, raw, unprepared	[H] sorghum, raw
[Y] yam, raw	[Z] plantains, raw
[I] rice, brown, long-grain, raw	

Arsenic concerns

Rice and rice products contain arsenic, a known poison and Group 1 carcinogen. There is no safe level of arsenic, but, as of 2012, a limit of 10 parts per billion has been established in the United States for drinking water, twice the level of 5 parts per billion originally proposed by the EPA. Consumption of one serving of some varieties of rice gives more exposure to arsenic than consumption of 1 liter of water that contains 5 parts per billion arsenic; however, the amount of arsenic in rice varies widely with the greatest concentration in brown rice and rice grown on land formerly used to grow cotton; in the United States, Arkansas, Louisiana, Missouri, and Texas. The U.S. Food and Drug Administration (FDA) is studying this issue, but has not established a limit. China has set a limit of 150 ppb for arsenic in rice.

White rice grown in Arkansas, Louisiana, Missouri, and Texas, which account for 76 percent of American-produced rice had higher levels of arsenic than other regions of the world studied, possibly because of past use of arsenic-based pesticides to control cotton weevils. Jasmine rice from Thailand and Basmati rice from Pakistan and India contain the least arsenic among rice varieties in one study.

Bacillus cereus

Cooked rice can contain Bacillus cereus spores, which produce an emetic toxin when left at 4–60 °C (39–140 °F). When storing cooked rice for use the next day, rapid cooling is advised to reduce the risk of toxin production. One of the enterotoxins produced by Bacillus cereus is heat-resistant; reheating contaminated rice kills the bacteria, but does not destroy the toxin already present.

RICE-GROWING ENVIRONMENTS

Rice can be grown in different environments, depending upon water availability. Generally, rice does not thrive in a waterlogged area, yet it can survive and grow hereinand it can also survive flooding.

1. Lowland, rainfed, which is drought prone, favors medium depth; waterlogged, submergence, and flood prone
2. Lowland, irrigated, grown in both the wet season and the dry season
3. Deep water or floating rice

4. Coastal Wetland
5. Upland rice is also known as Ghaiya rice, well known for its drought tolerance

HISTORY OF DOMESTICATION AND CULTIVATION

There have been plenty of debates on the origins of the domesticated rice. Genetic evidence published in the Proceedings of the National Academy of Sciences of the United States of America (PNAS) shows that all forms of Asian rice, both indica and japonica, spring from a single domestication that occurred 8,200–13,500 years ago in China of the wild rice Oryza rufipogon. A 2012 study published inNature, through a map of rice genome variation, indicated that the domestication of rice occurred in the Pearl River valley region of Chinabased on the genetic evidence. From East Asia, rice was spread to South and Southeast Asia. Before this research, the commonly accepted view, based on archaeological evidence, is that rice was first domesticated in the region of the Yangtze River valley in China.

Morphological studies of rice phytoliths from the Diaotonghuan archaeological site clearly show the transition from the collection of wild rice to the cultivation of domesticated rice. The large number of wild rice phytoliths at the Diaotonghuan level dating from 12,000–11,000BP indicates that wild rice collection was part of the local means of subsistence. Changes in the morphology of Diaotonghuan phytoliths dating from 10,000–8,000 BP show that rice had by this time been domesticated. Soon afterwards the two major varieties of indica and japonica rice were being grown in Central China. In the late 3rd millennium BC, there was a rapid expansion of rice cultivation into mainland Southeast Asia and westwards across India and Nepal.

In 2003, Korean archaeologists claimed to have discovered the world's oldest domesticated rice. Their 15,000-year-old age challenges the accepted view that rice cultivation originated in China about 12,000 years ago. These findings were received by academia with strong skepticism, and the results and their publicizing has been cited as being driven by a combination of nationalist and regional interests. In 2011, a combined effort by the Stanford University, New York University, Washington University in St. Louis, and Purdue University has provided the strongest evidence yet that there is only one single origin of domesticated rice, in the Yangtze Valley of China.

Rice spread to the Middle East where, according to Zohary and Hopf (2000, p. 91), O. sativa was recovered from a grave at Susa in Iran (dated to the 1st century AD).

REGIONAL HISTORY

In a recent study, scientist have found a link for differences in human culture based on either wheat or rice cultivating races since ancient times.

Africa

Fig. Rice crop in Madagascar

African rice has been cultivated for 3500 years. Between 1500 and 800 BC, Oryza glaberrima propagated from its original centre, theNiger River delta, and extended to Senegal. However, it never developed far from its original region. Its cultivation even declined in favour of the Asian species, which was introduced to East Africa early in the common era and spread westward. African rice helped Africa conquer its famine of 1203.

Asia

Fig. Indian women separating rice from straw

Fig. Cambodian women planting rice.

Today, the majority of all rice produced comes from China, India, Indonesia, Bangladesh, Vietnam, Thailand, Myanmar, Pakistan, Philippines, Korea and Japan. Asian farmers still account for 87% of the world's total rice production.

Sri Lanka

Rice is the staple food amongst all the ethnic groups in Sri Lanka. Agriculture in Sri Lanka mainly depends on the rice cultivation. Rice production is acutely dependent on rainfall and government supply necessity of water through irrigation channels throughout the cultivation seasons. The principal cultivation season, known as "Maha", is from October to March and the subsidiary cultivation season, known as "Yala", is from April to September. During Maha season, there is usually enough water to sustain the cultivation of all rice fields, nevertheless in Yala season there is only enough water for cultivation of half of the land extent.

Traditional rice varieties are now making a comeback with the recent interest in green foods.

Thailand

Rice is the main export of Thailand, especially the white jasmine rice 105 (Dok Mali 105). Thailand has a large number of rice varieties, 3,500 kinds with different characters, and 5 kinds of wild rice cultivates. In each region of the country there are different rice seed types. Their use depends on weather, atmosphere, and topography.

The northern region has both low lands and high lands. The farmers' usual crop is non-glutinous rice such as Niew Sun Pah Tong rice seeds. This rice is naturally protected from leaf disease, and the paddy has a brown color. The northeastern region has a large area, where farmers can cultivate about 36 million square meters of rice. Although most of them are plains and dry areas, they can grow the white jasmine rice 105 which is the most famous Thai rice. The white jasmine rice was developed in Chonburi province first and after that it was grown in many areas in the country but the rice from this region has a high quality, because it's softer, whiter and more fragrant.This rice can resist drought, acidic soil, and alkaline soil. The central region is mostly composed of plains. Most farmers grow Jao rice. For example, the Pathum Thani 1 rice which has qualities similar to the white jasmine 105 rice. Their paddy has the color of thatch and their cooked rice has fragrant grains also.

In the southern region, most farmers transplant around boundaries to the flood of plain or plain between mountains. Farming is the region is slower than other regions because the rainy season comes late. The popular rice varieties in this area are the Leb Nok Pattani seeds, a type of Jao rice. Their paddy has the color of thatch and it can be processed to make noodles.

Companion plant

One of the earliest known examples of companion planting is the growing of rice with Azolla, the mosquito fern, which covers the top of a fresh rice paddy's water, blocking out any competing plants, as well as fixing nitrogen from the atmosphere for the rice to use. The rice is planted when it is tall enough to poke out above the azolla. This method has been used for at least a thousand years.

Middle East

Rice was grown in some areas of Mesopotamia (southern Iraq). With the rise of Islam it moved north to Nisibin, the southern shores of the Caspian Sea (in Gilan and Mazanderan provinces of Iran) and then beyond the Muslim world into the valley of the Volga. In Egypt, rice is mainly grown in the Nile Delta. In Israel, rice came to be grown in the Jordan Valley. Rice is also grown in Saudi Arabia at Al-Hasa Oasis and in Yemen.

Europe

Rice was known to the Classical world, being imported from Egypt, and perhaps west Asia. It was known to Greece by returning soldiers from Alexander the Great's military expedition to Asia. Large deposits of rice from the first century A.D. have been found in Roman camps in Germany.

The Moors brought Asiatic rice to the Iberian Peninsula in the 10th century. Records indicate it was grown in Valencia and Majorca. In Majorca, rice cultivation seems to have stopped after the Christian conquest, although historians are not certain.

Muslims also brought rice to Sicily, where it was an important crop long before it is noted in the plain of Pisa (1468) or in the Lombard plain (1475), where its cultivation was promoted by Ludovico Sforza, Duke of Milan, and demonstrated in his model farms. After the 15th century, rice spread throughout Italy and then France, later propagating to all the continents during the age of European exploration.

In European Russia, a short-grain, starchy rice similar to the Italian varieties, has been grown in the Krasnodar Krai, and known in Russia as "Kuban Rice" or "Krasnodar Rice". In the Russian Far East several japonica cultivars are grown in Primorye around the Khanka lake. Increasing scale of rice production in the region has recently brought criticism towards growers' alleged bad practices in regards to the environment.

Caribbean and Latin America

Rice is not native to the Americas but was introduced to Latin America and the Caribbean by European colonizers at an early date with Spanish colonizers introducing Asian riceto Mexico in the 1520s at Veracruz and the

Portuguese and their African slaves introducing it at about the same time to colonial Brazil. Recent scholarship suggests thatenslaved Africans played an active role in the establishment of rice in the New World and that African rice was an important crop from an early period. Varieties of rice and bean dishes that were a staple dish along the peoples of West Africa remained a staple among their descendants subjected to slavery in the Spanish New World colonies, Brazil and elsewhere in the Americas.

The Native Americans of what is now the Eastern United States may have practiced extensive agriculture with forms of wild rice (Zizania palustris), which looks similar to but it not directly related to rice.

United States

Fig. South Carolina rice plantation, showing a winnowing barn (Mansfield Plantation, Georgetown)

In 1694, rice arrived in South Carolina, probably originating from Madagascar.

In the United States, colonial South Carolina and Georgia grew and amassed great wealth from the slave labor obtained from theSenegambia area of West Africa and from coastal Sierra Leone. At the port of Charleston, through which 40% of all American slave imports passed, slaves from this region of Africa brought the highest prices due to their prior knowledge of rice culture, which was put to use on the many rice plantations around Georgetown, Charleston, and Savannah.

From the enslaved Africans, plantation owners learned how to dyke the marshes and periodically flood the fields. At first the rice was laboriously milled by hand using large mortars and pestles made of wood, then winnowed in sweetgrass baskets (the making of which was another skill brought by slaves from Africa). The invention of the rice mill increased profitability of the crop, and the addition of water power for the mills in 1787 by millwright Jonathan Lucas was another step forward.

Rice culture in the southeastern U.S. became less profitable with the loss of slave labor after the American Civil War, and it finally died out just after the turn of the 20th century. Today, people can visit the only remaining rice plantation in South Carolina that still has the original winnowing barn and rice mill from the mid-19th century at the historic Mansfield Plantation in Georgetown, South Carolina. The predominant strain of rice in the Carolinas was from Africa and was known as 'Carolina Gold'. The cultivar has been preserved and there are current attempts to reintroduce it as a commercially grown crop.

In the southern United States, rice has been grown in southern Arkansas, Louisiana, and east Texas since the mid-19th century. Many Cajun farmers grew rice in wet marshes and low-lying prairies where they could also farm crayfish when the fields were flooded. In recent years rice production has risen in North America, especially in theMississippi embayment in the states of Arkansas and Mississippi.

Fig. Rice paddy fields just north of the city of Sacramento, California

Rice cultivation began in California during the California Gold Rush, when an estimated 40,000 Chinese laborers immigrated to the state and grew small amounts of the grain for their own consumption. However, commercial production began only in 1912 in the town of Richvale in Butte County. By 2006, California produced the second-largest rice crop in the United States, after Arkansas, with production concentrated in six counties north of Sacramento.

Unlike the Arkansas–Mississippi Delta region, California's production is dominated by short- and medium-grain japonica varieties, including cultivars developed for the local climate such as Calrose, which makes up as much as 85% of the state's crop.

References to wild rice in the Americas are to the unrelated Zizania palustris.

More than 100 varieties of rice are commercially produced primarily in six states (Arkansas, Texas, Louisiana, Mississippi, Missouri, and California) in the U.S. According to estimates for the 2006 crop year, rice production in the U.S. is valued at $1.88 billion, approximately half of which is expected to be exported. The U.S. provides about 12% of world rice trade. The majority of domestic utilization of U.S. rice is direct food use (58%), while 16% is used in each of processed foods and beer. 10% is found in pet food.

Australia

Rice was one of the earliest crops planted in Australia by British settlers, who had experience with rice plantations in the Americas and India.

Although attempts to grow rice in the well-watered north of Australia have been made for many years, they have consistently failed because of inherent iron and manganesetoxicities in the soils and destruction by pests.

In the 1920s it was seen as a possible irrigation crop on soils within the Murray-Darling Basin that were too heavy for the cultivation of fruit and too infertile for wheat.

Because irrigation water, despite the extremely low runoff of temperate Australia, was (and remains) very cheap, the growing of rice was taken up by agricultural groups over the following decades. Californian varieties of rice were found suitable for the climate in the Riverina, and the first mill opened at Leeton in 1951.

Even before this Australia's rice production greatly exceeded local needs, and rice exports to Japan have become a major source of foreign currency. Above-average rainfall from the 1950s to the middle 1990s encouraged the expansion of the Riverina rice industry, but its prodigious water use in a practically waterless region began to attract the attention of environmental scientists. These became severely concerned with declining flow in the Snowy River and the lower Murray River.

Although rice growing in Australia is highly profitable due to the cheapness of land, several recent years of severe drought have led many to call for its elimination because of its effects on extremely fragile aquatic ecosystems. The Australian rice industry is somewhat opportunistic, with the area planted varying significantly from season to season depending on water allocations in the Murray andMurrumbidgee irrigation regions.

PRODUCTION AND COMMERCE

Production

The world dedicated 162.3 million hectares in 2012 for rice cultivation and the total production was about 738.1 million tonnes. The average world farm yield for rice was 4.5 tonnes per hectare, in 2012.

Rice farms in Egypt were the most productive in 2012, with a nationwide average of 9.5 tonnes per hectare. Second place: Australia – 8.9 tonnes per hectare. Third place:USA – 8.3 tonnes per hectare.

Rice is a major food staple and a mainstay for the rural population and their food security. It is mainly cultivated by small farmers in holdings of less than 1 hectare. Rice is also a wage commodity for workers in the cash crop or non-agricultural sectors. Rice is vital for the nutrition of much of the population in Asia, as well as in Latin America and the Caribbean and in Africa; it is central to the food security of over half the world population. Developing countries account for 95% of the total production, with China and India alone responsible for nearly half of the world output.

World production of rice has risen steadily from about 200 million tonnes of paddy rice in 1960 to over 678 million tonnes in 2009. The three largest producers of rice in 2009 were China (197 million tonnes), India (131 Mt), and Indonesia (64 Mt). Among the six largest rice producers, the most productive farms for rice, in 2009, were in China producing 6.59 tonnes per hectare.

Many rice grain producing countries have significant losses post-harvest at the farm and because of poor roads, inadequate storage technologies, inefficient supply chains and farmer's inability to bring the produce into retail markets dominated by small shopkeepers. A World Bank – FAO study claims 8% to 26% of rice is lost in developing nations, on average, every year, because of post-harvest problems and poor infrastructure. Some sources claim the post-harvest losses to exceed 40%. Not only do these losses reduce food security in the world, the study claims that farmers in developing countries such as China, India and others lose approximately US$89 billion of income in preventable post-harvest farm losses, poor transport, the lack of proper storage and retail. One study claims that if these post-harvest grain losses could be eliminated with better infrastructure and retail network, in India alone enough food would be saved every year to feed 70 to 100 million people over a year. However, other writers have warned against dramatic assessments of post-harvest food losses, arguing that "worst-case scenarios" tend to be used rather than realistic averages and that in many cases the cost of avoiding losses exceeds the value of the food saved.

The seeds of the rice plant are first milled using a rice huller to remove the chaff (the outer husks of the grain). At this point in the process, the product is called brown rice. The milling may be continued, removing the bran, i.e., the rest of the husk and the germ, thereby creating white rice. White rice, which keeps longer, lacks some important nutrients; moreover, in a limited diet which does not supplement the rice, brown rice helps to prevent the disease beriberi.

Either by hand or in a rice polisher, white rice may be buffed with glucose or talc powder (often called polished rice, though this term may also refer to white rice in general), parboiled, or processed into flour. White rice may also be enriched by adding nutrients, especially those lost during the milling process. While the cheapest method of enriching involves adding a powdered blend of nutrients that will easily wash off (in the United States, rice which has been so treated requires a label warning against rinsing), more sophisticated methods apply nutrients directly to the grain, coating the grain with a water-insoluble substance which is resistant to washing.

In some countries, a popular form, parboiled rice, is subjected to a steaming or parboiling process while still a brown rice grain. This causes nutrients from the outer husk, especially thiamine, to move into the grain itself. The parboil process causes a gelatinisation of the starch in the grains. The grains become less brittle, and the color of the milled grain changes from white to yellow. The rice is then dried, and can then be milled as usual or used as brown rice. Milled parboiled rice is nutritionally superior to standard milled rice. Parboiled rice has an additional benefit in that it does not stick to the pan during cooking, as happens when cooking regular white rice. This type of rice is eaten in parts of India and countries of West Africa are also accustomed to consuming parboiled rice.

Despite the hypothetical health risks of talc (such as stomach cancer), talc-coated rice remains the norm in some countries due to its attractive shiny appearance, but it has been banned in some, and is no longer widely used in others (such as the United States). Even where talc is not used, glucose, starch, or other coatings may be used to improve the appearance of the grains.

Rice bran, called nuka in Japan, is a valuable commodity in Asia and is used for many daily needs. It is a moist, oily inner layer which is heated to produce oil. It is also used as a pickling bed in making rice bran pickles and takuan.

Raw rice may be ground into flour for many uses, including making many kinds of beverages, such as amazake, horchata, rice milk, and rice wine. Rice flour does not containgluten, so is suitable for people on a gluten-free diet. Rice may also be made into various types of noodles. Raw, wild, or brown rice may also be consumed by raw-foodist orfruitarians if soaked and sprouted (usually a week to 30 days – gaba rice).

Processed rice seeds must be boiled or steamed before eating. Boiled rice may be further fried in cooking oil or butter (known as fried rice), or beaten in a tub to make mochi.

Rice is a good source of protein and a staple food in many parts of the world, but it is not a complete protein: it does not contain all of the essential amino acids in sufficient amounts for good health, and should be combined with other sources of protein, such as nuts, seeds, beans, fish, or meat.

Rice, like other cereal grains, can be puffed (or popped). This process takes advantage of the grains' water content and typically involves heating grains in a special chamber. Further puffing is sometimes accomplished by processing puffed pellets in a low-pressure chamber. The ideal gas law means either lowering the local pressure or raising the water temperature results in an increase in volume prior to water evaporation, resulting in a puffy texture. Bulk raw rice density is about 0.9 g/cm^3. It decreases to less than one-tenth that when puffed.

Harvesting, drying and milling

Fig. Rice combine harvester Katori-city,Japan

Fig. After the harvest, rice straw is gathered in the traditional way from small paddy fields in Mae Wang District, Chiang Mai Province, Thailand

Unmilled rice, known as paddy (Indonesia and Malaysia: padi; Philippines, palay), is usually harvested when the grains have a moisture content of around 25%. In most Asian countries, where rice is almost entirely the product of smallholder agriculture, harvesting is carried out manually, although there is a growing interest in mechanical harvesting. Harvesting can be carried out by the farmers themselves, but is also frequently done by seasonal labor groups. Harvesting is followed by threshing, either immediately or within a day or two. Again, much threshing is still carried out by hand but there is an increasing use of mechanical threshers. Subsequently, paddy needs to be dried to bring down the moisture content to no more than 20% for milling.

A familiar sight in several Asian countries is paddy laid out to dry along roads. However, in most countries the bulk of drying of marketed paddy takes place in mills, with village-level drying being used for paddy to be consumed by farm families. Mills either sun dry or use mechanical driers or both. Drying has to be carried out quickly to avoid the formation of molds. Mills range from simple hullers, with a throughput of a couple of tonnes a day, that simply remove the outer husk, to enormous operations that can process 4,000 tonnes a day and produce highly polished rice. A good mill can achieve a paddy-to-rice conversion rate of up to 72% but smaller, inefficient mills often struggle to achieve 60%. These smaller mills often do not buy paddy and sell rice but only service farmers who want to mill their paddy for their own consumption.

Distribution

Because of the importance of rice to human nutrition and food security in Asia, the domestic rice markets tend to be subject to considerable state involvement. While the private sector plays a leading role in most countries, agencies such as BULOG in Indonesia, the NFA in the Philippines, VINAFOOD in Vietnam and the Food Corporation of India are all heavily involved in purchasing of paddy from farmers or rice from mills and in distributing rice to poorer people. BULOG and NFA monopolise rice imports into their countries while VINAFOOD controls all exports from Vietnam.

Trade

World trade figures are very different from those for production, as less than 8% of rice produced is traded internationally. In economic terms, the global rice trade was a small fraction of 1% of world mercantile trade. Many countries consider rice as a strategic food staple, and various governments subject its trade to a wide range of controls and interventions.

Developing countries are the main players in the world rice trade, accounting for 83% of exports and 85% of imports. While there are numerous importers of rice, the exporters of rice are limited. Just five countries – Thailand, Vietnam, China, the United States and India – in decreasing order of exported quantities, accounted for about three-quarters of world rice exports in 2002. However, this ranking has been rapidly changing in recent years. In 2010, the three largest exporters of rice, in decreasing order of quantity exported were Thailand, Vietnam and India. By 2012, India became the largest exporter of rice with a 100% increase in its exports on year-to-year basis, and Thailand slipped to third position. Together, Thailand, Vietnam and India accounted for nearly 70% of the world rice exports.

The primary variety exported by Thailand and Vietnam were Jasmine rice, while exports from India included aromatic Basmati variety. China, an exporter of rice in early 2000s, was a net importer of rice in 2010 and will become the largest net importer, surpassing Nigeria, in 2013. According to a USDA report, the world's largest exporters of rice in 2012 were India (9.75 million tonnes), Vietnam (7 million tonnes), Thailand (6.5 million tonnes), Pakistan (3.75 million tonnes) and the United States (3.5 million tonnes).

Major importers usually include Nigeria, Indonesia, Bangladesh, Saudi Arabia, Iran, Iraq, Malaysia, the Philippines, Brazil and some African and Persian Gulf countries. In common with other West African countries, Nigeria is actively promoting domestic production. However, its very heavy import duties (110%) open it to smuggling from neighboring countries. Parboiled rice is particularly popular in Nigeria. Although China and India are the two largest producers of rice in the world, both countries consume the majority of the rice produced domestically, leaving little to be traded internationally.

World's most productive rice farms and farmers

The average world yield for rice was 4.3 tonnes per hectare, in 2010.

Australian rice farms were the most productive in 2010, with a nationwide average of 10.8 tonnes per hectare.

Yuan Longping of China National Hybrid Rice Research and Development Center, China, set a world record for rice yield in 2010 at 19 tonnes per hectare on a demonstration plot. In 2011, this record was surpassed by an Indian farmer, Sumant Kumar, with 22.4 tonnes per hectare in Bihar. Both these farmers claim to have employed newly developed rice breeds and System of Rice Intensification (SRI), a recent innovation in rice farming. SRI is claimed to have set new national records in rice yields, within the last 10 years, in many countries. The claimed Chinese and Indian yields have yet to be demonstrated on seven-hectare lots and to be reproducible over two consecutive years on the same farm.

PRICE

In late 2007 to May 2008, the price of grains rose greatly due to droughts in major producing countries (particularly Australia), increased use of grains for animal feed and US subsidies for bio-fuel production. Although there was no shortage of rice on world markets this general upward trend in grain prices led to panic buying by consumers, government rice export bans (in particular, by Vietnam and India) and inflated import orders by the Philippines marketing board, the National Food Authority. This caused significant rises in rice prices. In late April 2008, prices hit 24 US cents a pound, twice the price of seven months earlier. Over the period of 2007 to 2013, the Chinese government has substantially increased the price it pays domestic farmers for their rice, rising to US$500 per metric ton by 2013. The 2013 price of rice originating from other southeast Asian countries was a comparably low US$350 per metric ton.

On April 30, 2008, Thailand announced plans for the creation of the Organisation of Rice Exporting Countries (OREC) with the intention that this should develop into a price-fixing cartel for rice. However, little progress had been made by mid-2011 to achieve this.

Worldwide consumption

As of 2009 world food consumption of rice was 531.6 million metric tons of paddy equivalent (354,603 of milled equivalent), while the far largest consumers were China consuming 156.3 million metric tons of paddy equivalent (29.4% of the world consumption) and India consuming 123.5 million metric tons of paddy equivalent (23.3% of the world consumption).Between 1961 and 2002, per capita consumption of rice increased by 40%.

Rice is the most important crop in Asia. In Cambodia, for example, 90% of the total agricultural area is used for rice production.

U.S. rice consumption has risen sharply over the past 25 years, fueled in part by commercial applications such as beer production. Almost one in five adult Americans now report eating at least half a serving of white or brown rice per day.

ENVIRONMENTAL IMPACTS

Fig. Work by the International Center for Tropical Agriculture to measure the greenhouse gas emissions of rice production.

Rice cultivation on wetland rice fields is thought to be responsible for 11% of the anthropogenic methane emissions. Rice requires slightly more water to produce than other grains. Rice production uses almost a third of Earth's fresh water.

Long-term flooding of rice fields cuts the soil off from atmospheric oxygen and causes anaerobic fermentation of organic matter in the soil. Methane production from rice cultivation contributes ~1.5% of anthropogenic greenhouse gases.Methane is twenty times more potent a greenhouse gas than carbon dioxide.

A 2010 study found that, as a result of rising temperatures and decreasing solar radiation during the later years of the 20th century, the rice yield growth rate has decreased in many parts of Asia, compared to what would have been observed had the temperature and solar radiation trends not occurred. The yield growth rate had fallen 10–20% at some locations. The study was based on records from 227 farms in Thailand, Vietnam, Nepal, India, China, Bangladesh, and Pakistan. The mechanism of this falling yield was not clear, but might involve increased respiration during warm nights, which expends energy without being able to photosynthesize.

RAINFALL

Temperature

Rice requires high temperature above 20 °C but not more than 35 to 40 °C. Optimum temperature is around 30 °C (Tmax) and 20 °C (Tmin).

Solar radiation

The amount of solar radiation received during 45 days after harvest determines final crop output.

Atmospheric water vapor

High water vapor content (in humid tropics) subjects unusual stress which favors the spread of fungal and bacterial diseases.

Wind

Light wind transports CO2 to the leaf canopy but strong wind cause severe damage and may lead to sterility (due to pollen dehydration, spikelet sterility, and abortive endosperms).

PESTS AND DISEASES

Rice pests are any organisms or microbes with the potential to reduce the yield or value of the rice crop (or of rice seeds). Rice pests include weeds, pathogens, insects, nematode, rodents, and birds.

A variety of factors can contribute to pest outbreaks, including climatic factors, improper irrigation, the overuse of insecticides and high rates ofnitrogen fertilizer application. Weather conditions also contribute to pest outbreaks. For example, rice gall midge and army worm outbreaks tend to follow periods of high rainfall early in the wet season, while thrips outbreaks are associated with drought.

Insects

Major rice insect pests include: the brown planthopper (BPH), several spp. of stemborers – including those in the genera Scirpophaga and Chilo, the rice gall midge,several spp. of rice bugs – notably in the genus Leptocorisa, the rice leafroller and rice weevils.

Diseases

Rice blast, caused by the fungus Magnaporthe grisea, is the most significant disease affecting rice cultivation. Other major rice diseases include: sheath blight, rice ragged stunt (vector: BPH), and tungro (vector: Nephotettix spp). There is also an ascomycete fungus, Cochliobolus miyabeanus, that causes brown spot disease in rice.

Nematodes

Several nematode species infect rice crops, causing diseases such as Ufra (Ditylenchus dipsaci), White tip disease (Aphelenchoide bessei), and root knot disease (Meloidogyne graminicola). Some nematode species such as Pratylenchus spp. are most dangerous in upland rice of all parts of the world. Rice root nematode (Hirschmanniella oryzae) is a migratory endoparasite which on higher inoculum levels will lead to complete destruction of a rice crop. Beyond being obligate parasites, they also decrease the vigor of plants and increase the plants' susceptibility to other pests and diseases.

Other pests

These include the apple snail Pomacea canaliculata, panicle rice mite, rats, and the weed Echinochloa crusgali.

Integrated Pest Management

Crop protection scientists are trying to develop rice pest management techniques which are sustainable. In other words, to manage crop pests in such a manner that future crop production is not threatened. Sustainable pest management is based on four principles: biodiversity, host plant resistance (HPR), landscape ecology, and hierarchies in a landscape – from biological to social. At present, rice pest management includes cultural techniques, pest-resistant rice varieties, and pesticides (which include insecticide). Increasingly, there is evidence that farmers' pesticide applications are often unnecessary, and even facilitate pest outbreaks. By reducing the populations of natural enemies of rice pests, misuse of insecticides can actually lead to pest outbreaks. The International Rice Research Institute (IRRI) demonstrated in 1993 that an 87.5% reduction in pesticide use can lead to an overall drop in pest numbers. IRRI also conducted two campaigns in 1994 and 2003, respectively, which discouraged insecticide misuse and smarter pest management in Vietnam.

Rice plants produce their own chemical defenses to protect themselves from pest attacks. Some synthetic chemicals, such as the herbicide 2,4-D, cause the plant to increase the production of certain defensive chemicals and thereby increase the plant's resistance to some types of pests. Conversely, other chemicals, such as the insecticide imidacloprid, can induce changes in the gene expression of the rice that cause the plant to become more susceptible to attacks by certain types of pests. 5-Alkylresorcinolsare chemicals that can also be found in rice.

Botanicals, so-called "natural pesticides", are used by some farmers in an attempt to control rice pests. Botanicals include extracts of leaves, or a mulch of the leaves themselves. Some upland rice farmers in Cambodia spread chopped leaves of the bitter bush (Chromolaena odorata) over the surface of fields after planting. This practice probably helps the soil retain moisture and thereby facilitates seed germination. Farmers also claim the leaves are a natural fertilizer and helps suppress weed and insect infestations.

Fig. Chloroxylon is used for Pest Management in Organic Rice Cultivation in Chhattisgarh, India

Among rice cultivars, there are differences in the responses to, and recovery from, pest damage. Many rice varieties have been selected for resistance to insect pests. Therefore, particular cultivars are recommended for areas prone to certain pest problems. The genetically based ability of a rice variety to withstand pest attacks is called resistance. Three main types of plant resistance to pests are recognized as nonpreference, antibiosis, and tolerance.

Nonpreference (or antixenosis) describes host plants which insects prefer to avoid; antibiosis is where insect survival is reduced after the ingestion of host tissue; and tolerance is the capacity of a plant to produce high yield or retain high quality despite insect infestation.

Over time, the use of pest-resistant rice varieties selects for pests that are able to overcome these mechanisms of resistance. When a rice variety is no longer able to resist pest infestations, resistance is said to have broken down. Rice varieties that can be widely grown for many years in the presence of pests and retain their ability to withstand the pests are said to have durable resistance. Mutants of popular rice varieties are regularly screened by plant breeders to discover new sources of durable resistance.

Parasitic weeds

Rice is parasitized by the weed eudicot Striga hermonthica, which is of local importance for this crop.

ECOTYPES AND CULTIVARS

Fig. Rice seed collection from IRRI

While most rice is bred for crop quality and productivity, there are varieties selected for characteristics such as texture, smell, and firmness. There are four major categories of rice worldwide: indica, japonica, aromatic and glutinous. The different varieties of rice are not considered interchangeable, either in food preparation or agriculture, so as a result, each major variety is a completely separate market from other varieties. It is common for one variety of rice to rise in price while another one drops in price. Rice cultivars also fall into groups according to environmental conditions, season of planting, and season of harvest, called ecotypes. Some major groups are the Japan-type (grown in Japan), "buly" and "tjereh" types (Indonesia); "aman" (main winter crop), "aus" ("aush", summer), and "boro" (spring) (Bengal and Assam). Cultivars exist that are adapted to deep flooding, and these are generally called "floating rice".

The largest collection of rice cultivars is at the International Rice Research Institute in the Philippines, with over 100,000 rice accessions held in the International Rice Genebank. Rice cultivars are often classified by their grain shapes and texture. For example, Thai Jasmine rice is long-grain and relatively less sticky, as some long-grain rice contains less amylopectin than short-grain cultivars. Chinese restaurants often serve long-grain as plain unseasoned steamed rice though short-grain rice is common as well. Japanese mochi rice and Chinese sticky rice are short-grain. Chinese people use sticky rice which is properly known as "glutinous rice" (note: glutinous refer to the glue-like characteristic of rice; does not refer to "gluten") to make zongzi. The Japanese table rice is a sticky, short-grain rice. Japanese sake rice is another kind as well.

Indian rice cultivars include long-grained and aromatic Basmati (??????) (grown in the North), long and medium-grained Patna rice, and in South India (Andhra Pradesh and Karnataka) short-grained Sona Masuri (also called as Bangaru theegalu). In the state of Tamil Nadu, the most prized cultivar is ponni which is primarily grown in the delta regions of the Kaveri River. Kaveri is also referred to as ponni in the South and the name reflects the geographic region where it is grown. In the Western Indian state of Maharashtra, a short grain variety called Ambemohar is very popular. This rice has a characteristic fragrance of Mango blossom.

Aromatic rices have definite aromas and flavors; the most noted cultivars are Thai fragrant rice, Basmati, Patna rice, Vietnamese fragrant rice, and a hybrid cultivar from America, sold under the trade name Texmati. Both Basmati and Texmati have a mild popcorn-like aroma and flavor. In Indonesia, there are also red and black cultivars.

High-yield cultivars of rice suitable for cultivation in Africa and other dry ecosystems, called the new rice for Africa (NERICA) cultivars, have been developed. It is hoped that their cultivation will improve food security in West Africa.

Draft genomes for the two most common rice cultivars, indica and japonica, were published in April 2002. Rice was chosen as a model organism for the biology of grasses because of its relatively small genome (~430 megabase pairs). Rice was the first crop with a complete genome sequence.

On December 16, 2002, the UN General Assembly declared the year 2004 the International Year of Rice. The declaration was sponsored by more than 40 countries.

BIOTECHNOLOGY

High-yielding varieties

The high-yielding varieties are a group of crops created intentionally during the Green Revolution to increase global food production. This project enabled labor markets in Asia to shift away from agriculture, and into industrial sectors. The first "Rice Car", IR8 was produced in 1966 at the International Rice Research Institute which is based in the Philippines at the University of the Philippines' Los Baños site. IR8 was created through a cross between an

Indonesian variety named "Peta" and a Chinese variety named "Dee Geo Woo Gen." Scientists have identified and cloned many genes involved in the gibberellin signaling pathway, including GAI1 (Gibberellin Insensitive) and SLR1 (Slender Rice).

Disruption of gibberellin signaling can lead to significantly reduced stem growth leading to a dwarf phenotype. Photosynthetic investment in the stem is reduced dramatically as the shorter plants are inherently more stable mechanically. Assimilates become redirected to grain production, amplifying in particular the effect of chemical fertilizers on commercial yield. In the presence of nitrogen fertilizers, and intensive crop management, these varieties increase their yield two to three times.

Future potential

As the UN Millennium Development project seeks to spread global economic development to Africa, the "Green Revolution" is cited as the model for economic development. With the intent of replicating the successful Asian boom in agronomic productivity, groups like the Earth Institute are doing research on African agricultural systems, hoping to increase productivity. An important way this can happen is the production of "New Rices for Africa" (NERICA). These rices, selected to tolerate the low input and harsh growing conditions of African agriculture, are produced by the African Rice Center, and billed as technology "from Africa, for Africa". The NERICA have appeared in The New York Times(October 10, 2007) and International Herald Tribune (October 9, 2007), trumpeted as miracle crops that will dramatically increase rice yield in Africa and enable an economic resurgence. Ongoing research in China to develop perennial rice could result in enhanced sustainability and food security.

Golden rice

Rice kernels do not contain vitamin A, so people who obtain most of their calories from rice are at risk of vitamin A deficiency. German and Swiss researchers have genetically engineered rice to produce beta-carotene, the precursor to vitamin A, in the rice kernel. The beta-carotene turns the processed (white) rice a "gold" color, hence the name "golden rice." The beta-carotene is converted to vitamin A in humans who consume the rice. Although some rice strains produce beta-carotene in the hull, no non-genetically engineered strains have been found that produce beta-carotene in the kernel, despite the testing of thousands of strains. Additional efforts are being made to improve the quantity and quality of other nutrients in golden rice.

The International Rice Research Institute is currently further developing and evaluating Golden Rice as a potential new way to help address vitamin A deficiency.

Expression of human proteins

Ventria Bioscience has genetically modified rice to express lactoferrin, lysozyme which are proteins usually found in breast milk, and human serum albumin, These proteins haveantiviral, antibacterial, and antifungal effects.

Rice containing these added proteins can be used as a component in oral rehydration solutions which are used to treat diarrheal diseases, thereby shortening their duration and reducing recurrence. Such supplements may also help reverse anemia.

Flood-tolerant rice

Flooding is an issue that many rice growers face, especially in South and South East Asia where flooding annually affects 20 million hectares. Standard rice varieties cannot withstand stagnant flooding of more than about a week, mainly as it disallows the plant access to necessary requirements such as sunlight and essential gas exchanges, inevitably leading to plants being unable to recover. In the past, this has led to massive losses in yields, such as in the Philippines, where in 2006, rice crops worth $65 million were lost to flooding.

In response to this hazard, a variety of rice named Swarna Sub1 was developed via marker-assisted selection, with the ability to withstand prolonged periods of around 14 days beneath a flooded plain. The submergence tolerance ability of this variety is conferred by the presence of the Sub1A gene, introgressed from the Indian cultivar FR13A into the flood-vulnerable (but high yielding) cultivar Swarna. Swarna Sub1 effectively enters a dormant, energy-conserving state upon being submerged in a flooded rice paddy, a process that involves the finely controlled metabolism of enzymes such amylases, starch phosphorylase and alcohol dehydrogenase, allowing the plant to survive with limited oxygen and sunlight unlike its standard variety relatives. Given that the presence of the Sub1A gene does not impact upon the quality or quantity of the rice obtained, this variety has been very popular, with 1.7 million hectares of land in India having Swarna Sub1 and other flood-resistant varieties used instead of conventional rice crops.

Drought-tolerant rice

Drought represents a significant environmental stress for rice production, with 19–23 million hectares of rainfed rice production in South and South East Asia often at risk.Under drought conditions, without sufficient water to afford them the ability to obtain the required levels of nutrients from the soil, conventional commercial rice varieties can be severely affected – for example, yield losses as high as 40% have affected some parts of India, with resulting losses of around US$800 million annually.

The International Rice Research Institute (IRRI) conducts research into developing drought-tolerant rice varieties, including the varieties 5411 and

Sookha dhan, currently being employed by farmers in the Philippines and Nepal respectively. In addition, in 2013 the Japanese National Institute for Agrobiological Sciences led a team which successfully inserted the DEEPER ROOTING 1 (DRO1) gene, from the Philippine upland rice variety Kinandang Patong, into the popular commercial rice variety IR64, giving rise to a far deeper root system in the resulting plants. This facilitates an improved ability for the rice plant to derive its required nutrients in times of drought via accessing deeper layers of soil, a feature demonstrated by trials which saw the IR64 + DRO1 rice yields drop by 10% under moderate drought conditions, compared to 60% for the unmodified IR64 variety.

Salt-tolerant rice

Soil salinity poses a major threat to rice crop productivity, particularly along low-lying coastal areas during the dry season. For example, roughly 1 million hectares of the coastal areas of Bangladesh are affected by saline soils. These high concentrations of salt can severely affect rice plants' normal physiology, especially during early stages of growth, and as such farmers are often forced to abandon these otherwise potentially usable areas.

Progress has been made, however, in developing rice varieties capable of tolerating such conditions; the hybrid created from the cross between the commercial rice variety IR56 and the wild rice species Oryza coarctata is one example. O. coarctata is capable of successful growth in soils with double the limit of salinity of normal varieties, but lacks the ability to produce edible rice. Developed by the International Rice Research Institute, the hybrid variety can utilise specialised leaf glands that allow for the removal of salt into the atmosphere. It was initially produced from one successful embryo out of 34,000 crosses between the two species; this was then backcrossed to IR56 with the aim of preserving the genes responsible for salt tolerance that were inherited from O. coarctata. Extensive trials are planned prior to the new variety being available to farmers by approximately 2017–18.

Environment-friendly rice

Producing rice in paddies is harmful for the environment due to the release of methane by methanogenic bacteria. These bacteria live in the anaerobic waterlogged soil, and live off nutrients released by rice roots. Researchers have recently reported in Nature that putting the barley gene SUSIBA2 into rice creates a shift in biomass production from root to shoot (above ground tissue becomes larger, while belowground tissue is reduced), decreasing the methanogen population, and resulting in a reduction of methane emissions of up to 97%. Apart from this environmental benefit, the modification also increases the amount of rice grains by 43%, which makes it useful tool in feeding a growing world population.

MEIOSIS AND DNA REPAIR

Rice is used as a model organism for investigating the molecular mechanisms of meiosis and DNA repair in higher plants. Meiosis is a key stage of the sexual cycle in which diploid cells in the ovule (female structure) and the anther (male structure) produce haploid cells that develop further into gametophytes and gametes. So far, 28 meiotic genes of rice have been characterized. Studies of rice gene OsRAD51C showed that this gene is necessary for homologous recombinational repair of DNA, particularly the accurate repair of DNA double-strand breaks during meiosis. Rice gene OsDMC1 was found to be essential for pairing of homologous chromosomes during meiosis, and rice gene OsMRE11 was found to be required for both synapsis of homologous chromosomes and repair of double-strand breaks during meiosis.

CULTURAL ROLES OF RICE

Rice plays an important role in certain religions and popular beliefs. In many cultures relatives will scatter rice during or towards the end of awedding ceremony in front of the bride and groom.

The pounded rice ritual is conducted during weddings in Nepal. The bride gives a leafplate full of pounded rice to the groom after he requests it politely from her.

In the Philippines rice wine, popularly known as tapuy, is used for important occasions such as weddings, rice harvesting ceremonies and other celebrations. Dewi Sri is the traditional rice goddess of the Javanese, Sundanese, and Balinese people in Indonesia. Most rituals involving Dewi Sri are associated with the mythical origin attributed to the rice plant, the staple food of the region. In Thailand a similar rice deity is known as Phosop; she is a deity more related to ancient local folklore than a goddess of a structured, mainstream religion. The same female rice deity is known as Po Ino Nogar in Cambodia and as Nang Khosop in Laos. Ritual offerings are made during the different stages of rice production to propitiate the Rice Goddess in the corresponding cultures.

GOLDEN RICE

Golden rice is a variety of rice (Oryza sativa) produced through genetic engineering to biosynthesize beta-carotene, a precursor ofvitamin A, in the edible parts of rice. The research was conducted with the goal of producing a fortified food to be grown and consumed in areas with a shortage of dietary vitamin A, a deficiency which is estimated to kill 670,000 children under the age of 5 each year.

Golden rice differs from its parental strain by the addition of three beta-carotene biosynthesis genes. The scientific details of the rice were first published in Science in 2000, the product of an eight-year project by Ingo Potrykus of the Swiss Federal Institute of Technologyand Peter Beyer of the University of Freiburg. At the time of publication, golden rice was considered a significant breakthrough in biotechnology, as the researchers had engineered an entire biosynthetic pathway. In 2005, a new variety called Golden Rice 2, which produces up to 23 times more beta-carotene than the original golden rice, was announced.

Fig. Golden rice (far) compared to white rice (near)

Although golden rice was developed as a humanitarian tool, it has met with significant opposition from environmental and anti-globalization activists. Studies have found that golden rice poses no risk to human health, and multiple field tests have taken place with no adverse side-effects to participants.

Golden Rice was one of seven winners of the 2015 Patents for Humanity Awards by the United States Patent and Trademark Office.

CREATION

Golden rice was designed to produce beta-carotene, a precursor of vitamin A, in the edible part of rice, the endosperm. The rice plant can naturally produce beta-carotene in its leaves, where it is involved in photosynthesis. However, the plant does not normally produce the pigment in the endosperm, where photosynthesis does not occur. A key breakthrough was the discovery that a single phytoene desaturase gene (bacterial CrtI) can be used to produce lycopene from phytoene in GM tomato, rather than having to introduce the multiple carotene desaturases that are normally used by higher plants. Lycopene is then cyclized to beta-carotene by the endogenous cyclase in Golden Rice.

Golden rice was created by transforming rice with only two beta-carotene biosynthesis genes:

1. psy (phytoene synthase) from daffodil (Narcissus pseudonarcissus)
2. crtI (carotene desaturase) from the soil bacterium Erwinia uredovora

(The insertion of a lcy (lycopene cyclase) gene was thought to be needed, but further research showed it is already being produced in wild-type rice endosperm.)

The psy and crtI genes were transformed into the rice nuclear genome and placed under the control of an endosperm-specific promoter, so they are only expressed in the endosperm. The exogenous lcy gene has a transit peptide sequence attached so it is targeted to theplastid, where geranylgeranyl diphosphate formation occurs. The bacterial crtI gene was an important inclusion to complete the pathway, since it can catalyze multiple steps in the synthesis of carotenoids up to lycopene, while these steps require more than one enzyme in plants. The end product of the engineered pathway is lycopene, but if the plant accumulated lycopene, the rice would be red. Recent analysis has shown the plant's endogenous enzymes process the lycopene to beta-carotene in the endosperm, giving the rice the distinctive yellow color for which it is named. The original golden rice was called SGR1, and under greenhouse conditions it produced 1.6 μg/g of carotenoids.

Subsequent development

Golden rice has been bred with local rice cultivars in the Philippines and Taiwan and with the American rice cultivar 'Cocodrie'. The first field trials of these golden rice cultivars were conducted by Louisiana State University Agricultural Center in 2004. Field testing provides a more accurate measurement of nutritional value and enables feeding tests to be performed. Preliminary results from the field tests have shown field-grown golden rice produces 4 to 5 times more beta-carotene than golden rice grown under greenhouse conditions. In 2005, a team of researchers at biotechnology company, Syngenta, produced a variety of golden rice called "Golden Rice 2". They combined the phytoene synthase gene frommaize with crtl from the original golden rice. Golden rice 2 produces 23 times more carotenoids than golden rice (up to 37 μg/g), and preferentially accumulates beta-carotene (up to 31 μg/g of the 37 μg/g of carotenoids). To receive the Recommended Dietary Allowance (RDA), it is estimated that 144 g of the most high-yielding strain would have to be eaten. Bioavailability of the carotene from golden rice has been confirmed and found to be an effective source of Vitamin A for humans.

In June 2005, researcher Peter Beyer received funding from the Bill and Melinda Gates Foundation to further improve golden rice by increasing the levels of or the bioavailability of pro-vitamin A, vitamin E, iron, and zinc, and to improve protein quality through genetic modification.

POTENTIAL USE TO COMBAT VITAMIN A DEFICIENCY

The research that led to golden rice was conducted with the goal of helping children who suffer from vitamin A deficiency (VAD). In 2005, 190 million children and 19 million pregnant women, in 122 countries, were estimated to be affected by VAD. VAD is responsible for 1–2 million deaths, 500,000 cases of irreversible blindness and millions of cases of xerophthalmia annually. Children and pregnant women are at highest risk. Vitamin A is supplemented orally and by injection in areas where the diet is deficient in vitamin A. As of 1999, there were 43 countries that had vitamin A supplementation programs for children under 5; in 10 of these countries, two high dose supplements are available per year, which, according to UNICEF, could effectively eliminate VAD.However, UNICEF and a number of NGOs involved in supplementation note more frequent low-dose supplementation should be a goal where feasible.

Because many children in countries where there is a dietary deficiency in vitamin A rely on rice as a staple food, the genetic modification to make rice produce the vitamin A precursor beta-carotene is seen as a simple and less expensive alternative to vitamin supplements or an increase in the consumption of green vegetables or animal products.

Initial analyses of the potential nutritional benefits of golden rice suggested consumption of golden rice would not eliminate the problems of vitamin A deficiency, but should be seen as a complement to other methods of vitamin A supplementation. Since then, improved strains of golden rice have been developed containing sufficient provitamin A to provide the entire dietary requirement of this nutrient to people who eat about 75g of golden rice per day. In particular, since carotenes are hydrophobic, there needs to be a sufficient amount of fat present in the diet for golden rice (or most other vitamin A supplements) to be able to alleviate vitamin A deficiency. In that respect, it is significant that vitamin A deficiency is rarely an isolated phenomenon, but usually coupled to a general lack of a balanced diet. The RDA levels accepted in developed countries are far in excess of the amounts needed to prevent blindness. Moreover, this claim referred to an early cultivar of golden rice; one bowl of the latest version provides 60% of RDA for healthy children.

RESEARCH

Dr. José L. Domingo of the Laboratory of Toxicology and Environmental Health, School of Medicine, at Rovira i Virgili University in Spain said, "According to the information reported by the WHO, genetically modified products that are currently on the international market have all passed risk assessments conducted by national authorities." These assessments found no risk to human health. Domingo advocates continued research in the areas of GM rice and its effects on humans.

Clinical trials/ food safety and nutrition research

In 2009, research results of a clinical trial of Golden Rice with adult volunteers from the USA were published in the American Journal of Clinical

Nutrition. It concluded that "beta carotene derived from Golden Rice is effectively converted to vitamin A in humans". In a summary about the research the American Society for Nutrition suggests the implications of the research are that "Golden Rice could probably supply 50% of the Recommended Dietary Allowance (RDA) of vitamin A from a very modest amount — perhaps a cup — of rice, if consumed daily. This amount is well within the consumption habits of most young children and their mothers".

It is well known that beta carotene is found and consumed in many nutritious foods eaten around the world, including fruits and vegetables. Beta carotene in food is a safe source of vitamin A.

The Food Allergy Resource and Research Program of the University of Nebraska undertook research in 2006 that showed the proteins from the new genes in Golden Rice did not show any allergenic properties.

In August 2012, Tufts University and others published new research on Golden Rice in the American Journal of Clinical Nutrition showing that the beta carotene produced by Golden Rice is as good as beta carotene in oil at providing vitamin A to children. The study states that "recruitment processes and protocol were approved", but questions have been raised about the use of Chinese children to test the effects of Golden Rice. In 2015 the journal retracted the study, because researchers acted unethically when providing Chinese children golden rice without their parents consent.

CONTROVERSY

Critics of genetically engineered crops have raised various concerns. An early issue was that golden rice originally did not have sufficient vitamin A. This problem was solved by the development of new strains of rice. The speed at which vitamin A degrades once the rice is harvested, and how much remains after cooking are contested. However, a 2009 study concluded that golden rice is effectively converted into vitamin A in humans and a 2012 study that fed 68 children ages 6 to 8 concluded that golden rice was as good as vitamin A supplements and better than the natural beta-carotene in spinach.

Greenpeace opposes the use of any patented genetically modified organisms in agriculture and opposes the cultivation of golden rice, claiming it will open the door to more widespread use of GMOs. The International Rice Research Institute (IRRI) has emphasised the non-commercial nature of their project, stating that "None of the companies listed... are involved in carrying out the research and development activities of IRRI or its partners in Golden Rice, and none of them will receive any royalty or payment from the marketing or selling of Golden Rice varieties developed by IRRI."

Vandana Shiva, an Indian anti-GMO activist, argued the problem was not the plant per se, but potential problems with poverty and loss of biodiversity. Shiva claimed these problems could be amplified by the corporate control of agriculture. By focusing on a narrow problem (vitamin A deficiency), Shiva argued, golden rice proponents were obscuring the limited availability of diverse and nutritionally adequate food. Other groups argued that a varied diet containing foods rich in beta carotene such as sweet potato, leaf vegetables and fruit would provide children with sufficient vitamin A. Keith West of Johns Hopkins Bloomberg School of Public Health has stated that foodstuffs containing vitamin A are often unavailable, only available at certain seasons, or too expensive for poor families in underdeveloped countries.

In 2008 WHO malnutrition expert Francesco Branca cited the lack of real-world studies and uncertainty about how many people will use golden rice, concluding "giving out supplements, fortifying existing foods with vitamin A, and teaching people to grow carrots or certain leafy vegetables are, for now, more promising ways to fight the problem".

In 2013, author Michael Pollan, who had critiqued the product in 2001, unimpressed by the benefits, expressed support for the continuation of the research.

Protests

On August 8, 2013 an experimental plot of golden rice being developed at IRRI in the Philippines was uprooted by protesters. Mark Lynas, a famous former anti-GMO activist, reported in Slate that the vandalism was carried out by a group of activists led by the extreme left-inclined Kilusang Magbubukid ng Pilipinas (KMP) (unofficial translation:Farmers' Movement of the Philippines), to the dismay of other protesters. No local farmers participated in the uprooting, only the small number of activists damaged the Golden Rice crops, because the farmers believe in their local customs which imply that killing a living rice plant is unlucky.

DISTRIBUTION

Potrykus has enabled golden rice to be distributed free to subsistence farmers. Free licenses for developing countries were granted quickly due to the positive publicity that golden rice received, particularly in Time magazine in July 2000. Monsanto Company was one of the first companies to grant free licences.

The cutoff between humanitarian and commercial use was set at US$10,000. Therefore, as long as a farmer or subsequent user of golden rice genetics does not make more than $10,000 per year, no royalties need to be paid. In addition, farmers are permitted to keep and replant seed.

PUFFED RICE

Puffed rice is a type of puffed grain made from rice, commonly used in breakfast cereal or snack foods, and served as a popular street food in India. It is usually made by heating rice kernels under high pressure in the presence of steam, though the method of manufacture varies widely.

Fig. Puffed Rice

Fig. Puffed rice for sale in Ulsoor Market,Bangalore.

Fig. spiced puffed rice

PRODUCTION

A traditional puffed rice called muri (sometimes spelled mouri) is made by heating rice in a sand-filled oven. Muri is to rice as popcorn is to corn. The processing involved makes rice less perishable. Mandakki is a staple food in many parts of Rayalaseema, North Karnataka,Odisha, Tripura, West Bengal and Bangladesh. Jhalmuri or Masalemandakki is a very popular preparation made from mandakki (muri).

Puffed rice is formed by the reaction of both starch and moisture when heated within the shell of the grain. Unlike popcorn, rice kernels are naturally lacking in moisture and must first be conditioned with steam. Puffed rice can be created by heating the steam-conditioned kernels either with oil or in an oven.

Rice puffed in this way is crisp, and known as "crisped rice". Oven-crisped rice is used to produce theRice Krispies breakfast cereal as well as the crisped rice used in Lion Bars, Nestlé Crunch, Krackel, and similar chocolate bars. Though not as dramatic a change when compared to popcorn, the process and end result are the same.

Another method of puffing rice is "gun puffing", where the grain is conditioned to the correct level of moisture and pressurised to around 200 PSI. When the pressure is suddenly released, the pressure stored inside the kernel causes it to puff out. This method produces a puffed rice which is spongy in texture.

Rice can also be puffed by making a rice dough, and extruding small pellets which are then rapidly heated. The moisture in the doughflash boils and puffs the rice up. A cereal such as Cap'n Crunch is extruded, cooked, cut, pressurized, puffed and dried in a continuous process.

Modern "Puffed Rice" is attributed to an American, Dr. Alexander P. Anderson who, stumbling across "puffing" while trying to ascertain the water content of a single granule of starch, introduced the first puffing machine at the World's Fair in Saint Louis, Missouri, in 1904. His eights "guns" that puffed grains for Fair goers were dubbed "The Eighth Wonder of the World" by an advertising billboard poster. Once the puffing principle, technique and technology had been discovered by Anderson, the competition to puff ready-to-eat American breakfast cereal took over the economy of Battle Creek, Michigan, with Kellogg's and Quaker Oats being two memorable and still active names to endure through the early puffing frenzy. After puffed breakfast cereal, came puffed rice cakes.

Health reform and emphasis on natural foods began in the late 1800s and was principally centered in Battle Creek. Two leaders were men with the well known name of Kellogg: Dr. John Harvey Kellogg and his brother Will Keith Kellogg, most notable for his idea of adding sugar to Dr. Kellogg's manufactured corn flakes.

USAGE

Puffed rice is an ingredient of bhel puri, a popular Indian chaat (snack). It is offered to Hindu gods and goddesses in all poojas in the South Indian states of Kerala and Tamil Nadu. Pilgrims of Sabarimala often pack puffed rice in their Irumudikettu along with jaggery meant to be offered to Lord Ayyappan. Tamil saints say that Lord Ganesh loves pori, so it should be offered to him without fail. Pori has been mentioned in various Tamil literatures as an offering to Hindu deities. Offerings of pori and jaggery made to Vinayagar (Lord Ganesh) are mentioned in the Tiruppugazh, a 15th-century anthology of Tamil religious songs, written by Tamil poet Arunagirinathar.

Pori production has been the main family business for centuries among many villages around Namakkal, Avinashi in Tamil Nadu. In Telangana, as a snack typically given to children, puffed rice or bongulu is made into ball with jaggery sugar syrup or bellam pakam.

Mudhi is a staple food of people of Odisha. northern Odisha, especially Baripada, Mayurbhanj district is significant for the production of Mudhi, where throughout the state it is eaten in breakfast. NGOs have taken forward initiatives to engage village women of northern Odisha for producing Mudhi. Intellectual property rights (IPR) Cell of Orissa University of Agriculture and Technology (OUAT) has decided to bring out Geographical indication (GI) registration of Mudhi.

Central Government of India in the banner of Make in India decided Mudhi from Odisha will be part Indian traditional food among 12 traditional dishes from different states that would be launched globally.

Puffed rice is referred to as mur-mure in some parts of India. In many parts of Andhra Pradesh, North Karnataka uggani along with Merapakai Bajji (Chilli Bajjis) are popular. In Karnataka, Mandakki Usli made from mandakki is also famous. In Mithila area, "murhi" is had with "kachari"-fried potato/onion chops, fried fish or with mutton curry. "Jhal-murhi" and "Murhi-Bhuja" are also very popular snacks in this area. In Madhya Pradesh, this is referred to as Parmal and its very often eaten with Sev as a snack and also used in Bhel.

In the United States and Europe, puffed rice is served with milk as a breakfast cereal, such as the brand Rice Krispies. Some chocolate bars, such as the Nestlé Crunch, include puffed rice, and puffed rice cakes are sold as low-calorie snacks.

RICE BRAN OIL

Rice bran oil is the oil extracted from the hard outer brown layer of rice after chaff (rice husk). It is notable for its high smoke pointof 232 °C (450 °F) and its mild flavor, making it suitable for high-temperature cooking methods such as stir frying and deep frying. It is popular as a cooking oil in several Asian countries, including Bangladesh, Japan, India and China.

USES

Rice bran wax, obtained from rice bran oil and palpanese extract, is used as a substitute for carnauba wax in cosmetics, confectionery, shoe creams and polishing compounds. It is an edible oil which is used in the preparation of vegetable ghee.

Composition

Rice bran oil has a composition similar to that of peanut oil, with 38% monounsaturated, 37% polyunsaturated, and 25% saturated fatty acids. The fatty acid composition is:

Fatty acid	Percentage
C14:0 Myristic acid	0.6%
C16:0 Palmitic acid	21.5%
C18:0 Stearic acid	2.9%
C18:1 Oleic acid (an Omega 9 fatty acid)	38.4%
C18:2 Linoleic acid (LA, an Omega 6 fatty acid)	34.4%
C18:3 α-Linolenic acid (ALA, an Omega 3 fatty acid)	2.2%

Physical Propreties of Crude & refined Rice bran oil

character	Crude Rice bran oil	Refined oil
Moisture	0.5-1.0%	0.1-0.15%
Density (15-15 °C)	0.913-0.920	0.913-0.920
Refractive Index	1.4672	1.4672
Iodine value	95-100	95-104
Saponification value	187	187
Unsaponifiable matter	4.5-5.5	1.8-2.5
Free fatty acids	5-15%	0.15-0.2%
oryzanol	2.0	1.5-1.8
Tocopherol	0.15	0.05
Color(Tintometer)	20Y+2.8R	10Y+1.0R

HEALTH BENEFITS

A component of rice bran oil is the antioxidant ?-oryzanol, at around 2% of crude oil content. Thought to be a single compound when initially isolated, it is now known to be a mixture of steryl and other triterpenyl esters of ferulic acids.

Also significant is the relatively high fractions of tocopherols and tocotrienols, together as vitamin E. Rice bran oil is also rich in other phytosterols.

Cholesterol

Literature review shows rice bran oil and its active constituents improve blood cholesterol by reducing total plasma cholesterol and triglycerides, and increasing the proportion of HDL cholesterol. Results of an animal study indicated a 42% decrease in total cholesterol with a 62% drop in LDL cholesterol, when researchers supplemented test subjects' diets with fractionated vitamin E obtained from rice bran oil.

Menopause

One small-scale study of ?-oryzanol, a mixture of chemicals found in rice bran oil, found that 90% of the women had some form of relief from hot flashes after taking a supplement of the purified concentrate for four to six weeks.

Antioxidant stability

The oryzanol content of the pan heated rice bran oil samples remains approximately the same even when heated at 180°C for 8 hours, while a decrease in oryzanol content was reported in the case of microwave heating at the same conditions.

Skin

Squalene is a compound present in Rice Bran Oil which is easily absorbed by the skin and keeps it soft, supple and smooth.

Anti Cancer

It is rich in Vitamin E which is powerful antioxidant and has antimutagenic properties which prevent from cancer. Vitamin E also helps in boosting your immunity.

Omega fatty acids and inflammation

Rice bran oil has more than 2% omega 3 fatty acids in it (more than olive oil) and a good amount of omega 6 too.

Rice bran oil has been tested to reduce cholesterol levels.

Rice Bran oil is anti-inflammatory too and some studies have shown that its consumption can reduce the effects of menopause like hot flashes

Calcium Absorption

Rice bran might help lower cholesterol because the oil it contains has substances that might decrease cholesterol absorption and increase cholesterol elimination. One of the substances in rice bran might decrease calcium absorption; this might help reduce the formation of certain types of kidney stones.

CONCLUSION ON RICE BRAN OIL

Rice Bran Oil is the healthiest oil according to the recommended most fatty acid levels by Indian Council of Medical Research.

The National Institute of Nutrition and The Indian Council of Medical Research recommend oils that have an equal proportion of saturated fatty acids, monounsaturated fats and polyunsaturated fats. Rice bran oil has an almost balanced fatty acid composition that is close to this ratio. Rice bran oil is rich in monounsaturated and polyunsaturated fats and free of trans-fats.

A human clinical study was conducted by scientists of National Institute of Nutrition, Indian Council of Medical Research to prove hypolipidemic effect of RBO. They studied this on 12 hyperlipidemic patients, Subjects were told to replace their usual oil (any oil used previously) with Rice Bran oil. A significant decrease in bad cholesterol was noticed after period of 15 days in all 12.

National Institute of Nutrition in its Dietary guidelines for year 2011 has mentioned that Oryzanol found in Rice Bran Oil is helpful in reducing cholesterol and oxidative damage due to ageing, inflammation which occur in chronic diseases. It is not just the MUFA: PUFA ratio that is important but the ratio of SFA: MUFA: PUFA that is important. And according to the latest recommendations by National Institute of Nutrition (NIN) and Indian Council of Medical Research (ICMR)the ideal fatty acid composition is 27-33%: 33-40%: 27-33%. And the fatty acid composition of Rice bran oil comes closest to these recommendations with the percentages at 24: 42: 34." The SFA: MUFA: PUFA ratios of various oils were discussed and it concluded with the fact that Rice Bran Oil indeed has the most ideal fat composition, better frying stability and offers unique health benefits due to phytosterols present in it.

3

Research and Risk-assessment Techniques for Improved Weed Management

INTRODUCTION

A number of reasons exist for studying weed seed banks. Perhaps Mayor and Dessaint (1998, pp. 95-96) best summarized these for agronomic purposes in the following statement:

- Seed banks are of ecological and evolutionary importance in the dynamics of weed populations and communities.
- Seed longevity and carryover of viable seeds in the soil from previous years can buffer the effects of weed control and hence maintain the weed problem. Some researchers have found that, with the exception of weeds with large seeds, the seed bank is a better indicator of long-term influences of agronomic practices on weeds than aboveground vegetation.

Seed banks are also studied for the purpose of anticipating forthcoming weed problems, assessing biodiversity and granivore food resources, and so forth. Although the goals of seed bank studies can be quite diverse, a common denominator among seed bank researchers typically involves an interest in sampling adequacy. Indeed, researchers new to the field often express anxiety regarding methodologies for sampling and quantifying seed banks. The purpose of this chapter is to provide some guidelines regarding sampling protocols for estimating weed seed banks in soil.

The recent symposium volume edited by Champion et al. (1998) provides one of the best available condensations of seed bank studies and methodologies. We encourage researchers who desire to begin working on seed banks to consult this book. Another very useful treatise on seed banks is Leck et al. (1989). The review by Roberts (1981) and chapters in Baskin and Baskin (1998) also contain much useful information. Otherwise, the first and most important guidance we can provide is that the objectives of the overall experiment should be matched to the need to have seed bank information. Will knowledge of seed bank composition and density provide insights that are more useful than those gained from knowledge of aboveground vegetation? Although seed bank analyses usually are not very expensive, they are labour-intensive. Consequently, the objectives of seed bank studies should be clear and unequivocal. Researchers must remember that the seed bank is part of a dynamic soil-plant-animal-microbe system, and the tedious work required to characterize the seed bank only provides us with a 'shapshot' in time.

The second most important guidance we can supply is that there is no universal sampling protocol applicable to all studies of soil seed banks. Each investigator has specific objectives and unique limitations in terms of labour and equipment. Moreover, each agro-ecosystem to be studied also has characteristics that may demand distinctive experimental protocols. Unique protocols are most obvious for agricultural systems with specific weeds: e.g. rare - vs common, small - vs large-seeded, and widely dispersed - vs aggregated species. However, the physical environment also plays a large role in sampling efficiencies and protocols. For instance, wet clay soils are much more difficult to sample than moist loam soils. These differences must be taken into account when designing practical protocols for assessing weed seed banks.

Seed bank researchers must devise protocols that are suitable to their objectives, equipment and labour constraints, and the agricultural system in which they work. Consequently, our goals in this report are to provide guidelines that may help researchers, especially those new to seed bank analyses and to tailor new studies on weed seed banks with a minimum of effort devoted to protocol development. This is not to suggest that new and more efficient protocols should not be devised. Instead, we urge new seed bank researchers not to duplicate and repeat weak protocols that could have been stronger with only slight modifications. These modifications often require no extra labour or expense. In some cases, only the time or type of sampling would need to be altered.

SAMPLING FIELD SOILS

Seed banks typically are confined to the surface and upper 30 cm of soil, although some perennial plants maintain seeds in aboveground seed banks (e.g. in serotinous cones of Pinus contorta Douglas ex Loudon). Therefore, sampling soil usually is a necessary component of seed bank studies. The most obvious questions that arise are: How many and what size soil samples should be taken?

The horizontal distribution of seeds across soil determines, in part, how many soil samples need to be taken. Weed seeds typically are not distributed randomly across a field. If they were, sampling seed banks would be much easier. Instead, weed seed banks almost always are highly aggregated in agricultural fields (Wiles and Schweizer, 1999, Chauvel et al. 1989). The aggregation can be the result of very limited dispersal away from parental plants, such as with early-maturing weeds in late-maturing crops (e.g. Avena fatua L. in soybean); or human-mediated dispersal of weeds that mature synchronously with crops, wherein seeds are spread in strips across fields by combine harvesters (e.g. A. fatua in wheat). Such aggregation affects the results of soil seed bank sampling.

The spatial pattern of seed banks often can be described mathematically by a negative binomial distribution (NBD). From a practical point of view, this basically means that many soil samples representative of the seed bank for any particular species will have no seeds, and a few samples will have high numbers of seeds. For instance, Jones, (1998) found that at least half of her sampled cores were devoid of seeds when mean seed densities were less than 4 000 m-2, and that 75 percent of cores were devoid of seeds when seed densities were less than 750 m-2.

Note that seed density changes the apparent level of aggregation. Typically, as the density of a species increases, the level of aggregation decreases, and the ease of adequate sampling rises, but much variation surrounds this generality. For instance, in the equation describing NBD, the level of aggregation is associated with the k coefficient, which was reported by Chauvel et al. (1989) for several common temperate weed species.

Table. Descriptive statistics for the distribution of seeds of five of the more abundant species across a field in France, where N is the total number of seeds detected in soil cores (4.7 cm diameter and 30 cm deep), m is mean seed number per core, CV is the coefficient of variation (s/m), and k is the NBD aggregation coefficient.

Species	*N*	*m*	CV	*k*
Thlaspi arvense L.	1105	5.58	0.72	2.79
Sinapis arvensis L.	1079	5.45	1.01	1.20
Chenopodium album L.	881	4.45	0.96	1.42
Alopecurus myosuroides Huds.	527	2.66	1.31	0.75
Fallopia convolulus (L.) Loeve	409	2.07	0.83	4.76

The most common species (highest value of m) was T. arvense, which had the lowest CV and one of the highest k values, meaning that it was not as aggregated as most other species.

The least common species was F. convolvulus, which might have been expected to be the most aggregated, but in this instance it was the least aggregated plant (highest k value). A. myosuroides was the most aggregated species (k was lowest), and its CV was highest and m almost the lowest among the species. These results suggest that the most difficult species to detect in this particular field would be A. myosuroides. Nevertheless, with the proper sample number and size, even the density of A. myosuroides could be established with some certainty.

How much soil to sample?

The initial question that needs to be answered is: "How much soil do I need to sample to get an accurate portrayal of the seed bank?" The amount of soil sampled is a product of the number of cores and the size of the cores. Core size involves core area or diameter (i.e. most soil sampling tools are tubes with circular orifices) and also core depth.

How many cores?

The absence of randomly varying seed populations in soil introduces problems for sampling seed banks. The primary question is how many soil cores should be sampled for an adequate representation of a seed bank. One of the best papers in recent years to address this issue was the synthesis of a multi-nation and multi-year study sponsored by the European Weed Research Society.

Abundant empirical evidence of seed bank densities from five countries showed a consistent relationship between mean seed bank density and variance. Overall, the relationship was defined as log10s2 = 0.45 + 1.41 log10 m, which is an adaptation of Taylor's Power Law. From this relationship, Dessaint et al. (1996) derived an equation that helps approximate sampling adequacy based upon differing levels of desired precision. That equation was N = 100.45 (m/509)-0.59 D-2

Where N is the estimated number of necessary samples (i.e. 5-cm diameter soil cores) to adequately represent a seed bank, and D represents the desired level of precision. D is defined as the standard error of the mean divided by the mean (SEm/m).

The value of m is divided by 509 to convert the area of a 5-cm diameter core to 1 m2. Dessaint et al. (1996) indicated that a D value of 0.3 was a practical level of precision for seed bank studies. We believe that even a precision value of 0.5, which is less precise than 0.3, may be adequate depending upon the goals of the researchers. For instance, perhaps D could be set to 0.4-0.5 for species that are relatively uncommon but easily controlled. In contrast, species that are both common and difficult to manage probably merit D values of 0.2-0.3. Thus, sampling efforts can be conditioned by the required value of the resulting information.

For hypothetical seed densities of 10, 50, 100, 500, 1000, 5 000, and 10 000 seeds m-2, each with a precision level of 0.2, 0.3, 0.4, and 0.5. These results by Dessaint et al. (1996) are dependent upon the use of soil cores that are 5 cm in diameter. Results would differ for cores with other diameters: higher numbers of samples for smaller diameter cores, and fewer samples for larger cores. However, as stressed below, 5-cm diameter cores are an ideal size for seed bank studies, and we advocate the use of this size of soil sampling tool.

Table. The number of soil cores (5 cm diameter) necessary to determine seed bank densities under four levels of desired precision assuming various seed densities.

Seed bank (seeds m^{-2})	Precision level (*D*)			
	0.2	0.3	0.4	0.5
10	716	318	179	115
50	277	123	69	44
100	184	82	46	29
500	71	32	18	11
1000	47	21	12	8
5000	18	8	5	3
10000	12	5	3	2

The specific to the region in which the data were collected, but this region was quite broad, involved many crops and soil types, crossed several national boundaries, and spanned Mediterranean through temperate agro-ecosystems. In any event, the number of cores necessary for estimating seed bank densities is not nearly as large as other literature sources indicate, provided that densities of the species of interest are greater than 100 seeds m-2. With some luck, these results will have universal application (although exceptions certainly can be expected), which will help to alleviate a formidable burden on seed bank researchers.

It should be kept in mind, however, that if the goal of the study is to characterize the seed flora and density completely, as in a weed community analysis, then the number of cores required is higher because the less common species will be sampled at a lower level of precision than the more common species. In agricultural fields, over 90 percent of the seeds might be comprised of just a few of the 30 or more species represented in the seed bank. Therefore, some of the more interesting changes in weed communities - such as shifts in response to management - might be happening among the less abundant species, and higher sampling intensity is required to detect these changes.

Although seed banks range greatly in density, the median value for crop fields in Minnesota is about 1000 seeds m-2. Estimation of this density would require 21 samples for the recommended precision level of 0.3. Clearly, species with very low densities (<100 seeds m-2) would require so many soil cores that precise determination of their seed banks is not practical. An example of such a species is Xanthium strumarium L.

Its seeds are dispersed in two-seeded capsules (burs), the size of which is about 1 × 2 cm. Not many of these large fruits are produced relative to other weeds with smaller seeds. Consequently, detection of X. strumarium in seed banks is rare, except in very dense infestations, so that an accurate estimation of the size of its seed bank is quite difficult. Despite its importance as a weed and the desire of researchers to understand its seed bank dynamics, such a large-seeded species is not easily amenable to seed bank analysis.

What size core?

The diameter of the core usually depends upon available equipment. Most hand-held soil sampling equipment was developed for soil scientists, and much of it is about 2-3 cm in diameter. Although, theoretically, any soil core diameter is suitable for sampling weed seed banks, certain sizes are far more practical than others.

As mentioned above, fewer large-diameter cores than small-diameter cores are necessary to sample seed banks adequately. However, large diameter cores amass great quantities of soil quickly and may overwhelm the researcher. For example, a single 10 cm diameter soil core (15 cm depth) has a dry weight of about 1-2 kg. Hypothetically, even if only ten cores were exhumed per plot, the total weight (soil plus soil water) of the sample might be 20 kg. If an experiment had ten treatments and five replications, the total mass of sampled soil might be as much as 1 000 kg. Such massive amounts of soil can be unwieldy for transportation to and from the field site, as well as within the laboratory.

In contrast, the soil in narrow cores (e.g. 2 cm diameter) might weigh only 50-100 g, which makes them easy to handle and transport. However, the probability of detecting seeds in such small amounts of soil is so very low that that many cores must be sampled for compensation. There are very few studies that compare core sizes for sampling efficiency. Benoit et al. (1989) found that augers with diameters of 1.9, 2.7, and 3.3 cm were not different in estimating the number of Chenopodium album seeds when a similar volume of soil was sampled. Although the largest and smallest of these diameters differ only by a factor of 1.7, the largest and smallest volumes of sampled soil from a single core exhumed by these augers differ by a factor of 3. In other words, three times as many soil cores of 1.9 cm diameter would need to be sampled as cores of 3.3 cm diameter in order to estimate similar seed bank densities. The additional labour involved in taking more cores may not compensate for the ease of using small-diameter cores.

Our experience is that 5 cm diameter cores represent a practical solution to the problem of core sizes. This size core is large enough to detect seeds, but small enough not to burden the researcher with too much soil. We advocate their use in seed bank studies.

However, also other factors are involved in the choice of diameters of soil-coring devices. The most significant of these is soil texture and soil water content. Wet soils with high percentages of expanding clays are notoriously difficult to remove from sample tubes, especially from small-diameter tubes. Cores with diameters of up to 10 cm should be considered for such soils.

Application of non-toxic oils (vegetable oils) to the coring implement helps greatly in preventing clay from sticking to the device. Orifices that are a few millimetres narrower than the diameter of the sampling tube also may aid in preventing the soil from adhering too tightly to the inside of the sample tube. However, these types of coring tools are more likely to compress soils with low bulk densities while the coring tool is being driven into the soil. This compaction of the core confounds reliability of core depths. In contrast, very dry soil can resist penetration by coring implements. In these cases, narrow cores may be more practical than the recommended 5 cm diameter cores. Researchers must be practical and balance multiple factors when choosing sampling equipment.

Vehicle-mounted soil probes are available and very useful for sampling large plots and fields. Hydraulics drive the probes into the soil, retract the probes, and remove soil from the probes, which greatly facilitates sampling in many types of conditions. However, vehicle-mounted equipment is clumsy in small research plots. Hand-held equipment often is easier to use in these cases.

What depth of soil to sample?

The depth to which soil cores should be taken is entirely dependent upon the objectives of the research. Generally, few seedlings have the ability to emerge if their seeds are buried deeper than 10 cm. Exceptions include large-seeded species such as A. fatua and X. strumarium. Consequently, soil samples rarely need to exceed 10 cm depth. However, many seed bank researchers are interested in differing tillage systems wherein seeds are buried differentially at depths by the tillage implements. In these cases, samples to 30 cm depth may be necessary.

These researchers should recall, however, that seed burial by mechanical tillage equipment is a physical process that is quite consistent. In other words, the same type of plough buries seeds in the same proportions and at the same depths regardless of soil type, location, time of year, etc. This consistency means that if the tillage system is known, then the relative proportions of seeds at different depths can be estimated without ever sampling the seed bank. Consequently, only a single depth may need to be sampled to estimate the seed bank of the entire soil profile, and that depth should probably be a function of the emergence-from-depth characteristics of the species of greatest concern to the researcher.

What spatial arrangement of samples?

Whether sampling soils within a plot or an entire field, researchers must decide on the spatial distribution of the samples. Random sampling designs would be appropriate if seeds were distributed randomly. Furthermore, the amount of time spent locating random points in a field, based upon a priori selection of random numbers, is usually not practical. For ease of sampling, many researchers take soil cores at roughly evenly-spaced intervals along a simple W-shaped pattern within a plot or field. Others have used X-shaped patterns or a single diagonal transect. Colbach et al. (2000) examined various patterns for sampling accuracy in aggregated weed populations and concluded that many patterns provided equivalent results, with the diagonal transect being the simplest design to use. We recommend that any design is acceptable provided that it somehow spans the length and width of the plot or field.

When to sample seed banks?

We cannot stress too greatly the need for logic when deciding the appropriate time of year to sample seed banks. Several authors report a lack of correspondence between aboveground vegetation and seed bank composition and density. The absence of such a relationship too often reflects the illogical times that these authors chose to sample the seed bank. For the sake of camaraderie we will not cite references in this regard, but even a casual perusal of Methods and Materials sections of the seed bank literature will confirm our assertion.

Basically, if a goal of the research is to relate seed banks to forthcoming aboveground vegetation, then seed banks should be sampled at times that follow seed shed but precede seed germination. Sampling seed banks after seedling emergence has little value, in theory or in practice. Samples need to be taken at a time that makes sense for the objectives of the study, based on the phenology of seed dispersal and germination in the habitat of interest.

Thus, in temperate zones, seed banks of summer annual weeds should be sampled prior to the first springtime flush of seedling emergence. Similarly, seed banks of winter annuals should be sampled before emergence of the first seedlings in autumn. Analogous logic should be used for sampling times in semi-tropical and tropical zones with distinct wet and dry seasons. The same reasoning should apply to irrigated land regardless of season; that is, the soils should be sampled before the onset of irrigation and subsequent seed germination.

Even when soil samples are taken before seed germination, there still can be a question regarding time of sampling. For example, in northern temperate zones, seeds of many summer annual weeds are shed during August through October and germinate during the following March through June, the exact times being species-dependent. The period of time for properly sampling seed banks of these species would be late autumn through early spring (e.g. November to March). Even during this winter period of apparent quiescence, however, there is biological activity with regard to buried seeds. This activity affects seed mortality, and it may influence the times for best sampling of seed banks.

Comparisons of viable seed banks of summer annual weeds from two sampling times (immediately after seed shed in autumn and immediately before seed germination in spring) indicated an approximate 10 percent loss in viability over winter and a slight superiority of springtime samples for predicting forthcoming aboveground weed densities (Forcella, 1992). Theoretically, this is exactly what would be expected. Aboveground vegetation of annual weeds should be a better reflection of the seed bank immediately before seed germination than the seed bank several months in the past because many and varied mortality events could have occurred during the intervening period.

Although the theory of seed bank sampling design may point to sampling immediately before seed germination, the practical aspects of seed bank research may justify sampling sooner (but still only after seed shed). Researchers need to balance desired accuracy with workloads. In the above example, spring not only is the best time to sample soils, but it is the only time to sow crops and implement many weed control procedures. Consequently, little time exists in spring for the extra work involved with sampling seed banks. Earlier sampling of seed banks can be justified for this reason alone, if not in conjunction with other reasons. Researchers often have to balance good protocols with labour constraints.

An additional reason for early sampling involves the length of time necessary to process soil samples in the laboratory or glasshouse. If the goal of the research is to use seed bank information to help make recommendations for weed management treatments then the information must be available at the time the treatments are to be implemented. This could be as early as days or even weeks before experimental plots are sown with crops in the case of early preplant herbicide treatments. Consequently, sampling soils months before crop sowing may be necessary.

HOW SHOULD SOIL CORES BE PROCESSED?

Once cores have been exhumed from the soil, there are two primary techniques for enumerating the number of seeds in these cores. The two methods give differing results, but the results of the two methods usually correlate with one another (Ball and Miller 1989, Barberi et al. 1998, Cardina and Sparrow, 1996; Forcella 1992).

Direct seed extraction

The first technique can be termed "direct seed extraction," and Malone, (1967) is the author most often cited for this technique. The direct seed extraction method can be used on (a) the entire soil sample derived from entire individual cores, (b) subsamples of individual cores, or (c) subsamples of soil from aggregated cores. Clearly, labour requirements decrease from a-c, as does reliability of the resulting estimates of seed bank densities.

A typical soil core of 5 cm diameter and 10 cm depth has a dry weight of about 200-300 g. Naturally, if labour is not in short supply, extracting seed from the entire soil core is preferred. However, a shortage of labour (or associated enthusiasm) is common in seed bank studies. Thus, some understanding of what proportion of a soil core must be examined is important. Analyses of differing amounts of well-mixed soil, in 20 g increments from typical cores indicated that, in general, 100 g was necessary for an adequate representation of the entire soil core (Forcella, 1992).

Subsampling soil from aggregated cores would be recommended most often in studies where individual cores were small (< 5 cm diameter), but where many cores were exhumed. Examining the entire soil volume from small-diameter cores typically would be futile because the probability of small cores harbouring even one seed is very low.

In the direct seed extraction technique, seeds are separated from soil typically by washing or flotation. The washing method has many variations. Most simply, the soil sample is placed on a screen with a mesh size smaller than the smallest expected seed. However, sieving risks loss of seeds that might be similar in size or shape to the objects being separated, or that might adhere to them. Sieving - especially for dry samples - can damage seeds that are thin, light, and fragile, but it can aid in scarifying seeds with hard seed-coats. Several researchers have used a series of sieves with different mesh sizes to sort seeds according to size. Mesh size is a critical factor in determining the efficiency of seed separation. A mesh size of about 0.2 mm can catch most small seeds, but would not be effective for dust-like seeds of species like Orbanche. There can also be considerable variation in seed size among seeds from a given species, even from a single plant. Therefore, the mesh size chosen to detect seeds of a given species must be small enough to catch the smallest individuals of that species.

The sample can be pre-soaked for a short time to saturate and loosen clay aggregates. Soaking the soil sample in a solution of sodium hexametaphosphate will improve dispersal of clay aggregates. The next step is to remove clay, silt, and fine sand particles from the sample. This commonly is done by shaking the sample while it is held by the screen, or by passing a jet of water over the sample.

Once the fine particles have passed through the screen, the remainder of the sample includes seeds, organic debris, sand particles, and in clay-rich soils, clay aggregates that did not fully disperse. These latter clay aggregates often can be eliminated by applying gentle pressure with fingertips until the aggregates crush and pass through the screen. The seeds and organic debris that remain on the screen are separated from the sand particles by differential flotation. If sand particles are not abundant, the sample can washed onto mesh (e.g. cheese cloth) and air-dried, whereupon the seeds are separated from the organic debris by hand. Air-driven seed cleaners also can be used to separate organic debris from seeds in dried samples.

An elutriator is a device that mechanically performs the same procedures as described in the preceding paragraph. The beauty of elutriators is that they can process several samples simultaneously. The equivalent of a primitive, but motorized, clothes washing machine was used successfully for removing soil from buried seed samples by Fay (1978). These machines are a convenience, but are not essential for seed bank analyses.

The flotation method often is used after the soil sample has been washed free of clay, silt, and fine sands, but whole unprocessed samples can be used as well. Here, the goal is to affect the buoyancy of seeds and soil particles differentially. A number of differing salts can be used for this purpose. Potassium carbonate has proven to be useful in this endeavour, in that it permits separation of the seeds from the soil particles. Short exposure to it is not toxic for seeds of some species, but it may have a detrimental effect on others. Some organic debris typically floats with the seeds. If large tubes are used to hold the samples and potassium carbonate solution, these can be centrifuged to affect separation of seeds from soil particles (Buhler and Maxwell, (1993)).

This method is most useful if a single species is of interest, and detergent and salt concentrations that are effective, but not toxic, can be determined. All direct seed extraction methods provide estimates of total seed bank densities, including densities of dead seeds.

Thus, this technique is especially valuable for studies involved with population dynamics of weeds. The technique may not be always appropriate for the correlation of seed banks with seedling populations, as the technique may confuse dead, dormant, and non-dormant seeds with one another. Additional and routine tests are available to determine viability in the isolated seeds, but as yet there is no routine method to distinguish dormant from non-dormant seeds.

Seed identification

Once seeds are more-or-less isolated using the direct seed extraction method, the seeds must then be identified. The sample material that remains after direct seed extraction typically is not pure seed, but a mixture of seeds, other organic materials, and soil particles. Perhaps the most time-consuming portion of the direct seed extraction method is examining these mixtures under magnification, and locating and then identifying the seeds. A thoroughly experienced and keenly-sighted researcher with a catalogued seed collection is the best possible tool for seed identification, but failing that, some excellent handbooks exist. Most of these texts are regional. Delorit, (1970) for example, is an excellent resource for North American researchers.

Image analysis, that is, computerized analysis of electronic images of isolated seeds, holds some promise for identifying weed seeds. However, little effort seems to have been devoted to this topic recently (Benoit et al. 1992, Buhler and Maxwell, 1993), probably because the human eye still can distinguish seeds and species so much more rapidly than any machine.

Another interesting and modern tool for weed seed identification is DNA fingerprinting. This method may be most appropriate for identifying species and biotypes with seeds that are indistinguishable visually (Fennimore et al. 1999, Joel et al. 1998, Mucher, 2000).

The hand-separation and counting processes generally are performed on samples that have dried after sieving and flotation. Unfortunately, seeds of some species (e.g. Impatiens spp.) lose viability quickly after drying, which introduces error in estimates of viable seed densities.

Viability testing

The seeds that are isolated through direct seed extraction may be viable or dead. These seeds can be tested for viability. The simplest viability test is to probe the seeds with fine-tipped forceps, remove obviously dead seeds, and then attempt to germinate those seeds that appear firm. The number of seeds that germinate provides an estimate of the abundance of seeds that are both viable and non-dormant. However, it may not provide information relevant to the number of viable but dormant seeds.

The viability of seeds can be determined through the well-known tetrazolium chloride (TZ) test. The TZ test is simple for the purposes of most weed researchers, but for seed technologists the test can be quite complex. Typically, seeds are soaked in a 0.1-0.2 percent TZ solution for a few hours to one week at 10-30 C, depending upon species and research objectives. The hydrogen released by dehydrogenase reactions in living tissues combines with TZ to form a red pigment. Thus, if seeds exposed to TZ eventually turn pink or red, they contain living tissue, whereas those without the red stain are presumed to be dead.

Complexity of the TZ test arises in many forms. For example, when the growing point (embryonic axis) within a seed is no longer able to grow, its cotyledons and other tissues may still contain enough dehydrogenase and produce sufficient H to elicit a pink/red response to TZ. Furthermore, microorganisms that consume dead and dying seeds also contain dehydrogeanses and produce H, which can react with TZ and produce elicit false positive results. Careful observation of seeds after exposure to TZ can eliminate these errors. Observations of red staining of the radicle and hypocotyl or coleoptile are critically important for proper determination of viability.

A publication of the Association of Official Seed Analysts (AOSA), Tetrazolium Testing Handbook (Peters, 2000) (http://www.aosaseed.com/

tetra/TZcommitteemain.html) provides many excellent species-specific drawings and insights for the proper use of TZ in the commercial seed industry. For weeds, however, exact determination of seed viability is more of a research interest than an economic and industrial requirement. Consequently, some of the highly structured and species-specific AOSA guidelines may be relaxed. The following two paragraphs describe procedures that we have found to be useful for some species common to North America.

Where possible, split air-dried seeds symmetrically with a single-edged razor to bisect and expose the embryo. Thus, the cut surface of each half of the seed should show at least parts of the radicle and hypocotyl (e.g. Abutilon theophrasti Medik.) or the radicle and coleoptile (e.g. Setaria faberi Herrm.). Choose the half-seed that appears most intact, place on blotter paper saturated with 0.2 percent TZ, and incubate at 25 C. After 12 hours' incubation, viable seeds exhibit red growing points, whereas dead seeds retain their original colour. The period of incubation usually is too short for substantial growth of micro-organisms.

Seeds of some species are not amenable to symmetrical bisection along the embryonic axis because of their small size or shape (e.g. Chenopodium album). These types of seeds can be split in any fashion with a sharp razor so that the radicle or hypocotyl is exposed on at least one half of the seed. Because the appropriate half-seed is difficult to discern at this stage, incubate both halves of the seed in TZ. The seed is viable if either half exhibits red after 12 hours' incubation.

Verification

The value of the data obtained can be improved by testing and calibration at several steps in the direct seed extraction process. The process can be validated by adding a known number of seeds to test samples to verify that they can be separated, identified, and counted with acceptable accuracy. If accuracy is low, determining which step in the process is limiting and making appropriate adjustments are important for correcting the protocol. Researchers need, for example, to ask the following: Are seeds lost during sieving? Can they be distinguished from other organic matter or soil mineral contaminants? Can viability be determined accurately? Do the extraction or floatation procedures affect viability? Seed identification and counting accuracy generally decrease with increased seed number, smaller seed size, and time-related fatigue of personnel involved.

Current counting methods for determining the density of viable seeds in seed bank samples are laborious and impractical as a means of characterizing the species composition of the seed bank. Worker exposure to salt solutions and detergents, as well as hours spent peering through a hand- lens or dissecting scope, are serious limitations. A method that works for some species is likely to be inadequate for other species whose seeds differ in shape, size, durability, ease of identification, and dormancy characteristics. Agricultural seed banks in the north and central United States often contain between 20-50 species, some of which occur in very low numbers. A single method is unlikely to accurately separate, detect, and correctly identify all species. Thus, the seed separation approach is only appropriate for one - or just a few - large-seeded target species where accuracy can be verified.

GERMINATION METHOD

The second major technique for enumerating seeds in the soil seed bank is referred to as the "germination" method. This technique is primarily used to enumerate the density of non-dormant seeds in the seed bank. In this case, soil cores typically are aggregated in logical units (e.g. 20 cores from a single plot that represents an experimental treatment). To aggregate cores, they are mixed thoroughly and then inserted in trays that are placed in incubators, greenhouse (glasshouse) benches, cold frames, or nurseries depending upon the objectives of the experiment and availability of facilities. Samples must be protected from seed contamination, disturbance, and granivores and herbivores. Protection is especially important in outdoor nurseries, but even within modern greenhouses, airborne seeds of species such as dandelion (Taraxicum officinale Weber) are common contaminants that enter through ventilation systems.

Soil depth in the trays should be no greater than the depth from which expected species can germinate, typically less than 5 cm, with 2-3 cm best for small-seeded species. If the soil is clay-rich, it can be mixed with known volumes of clean sand or commercial potting media to improve drainage. A useful procedure to improve drainage is to line the bottom of the tray with sand, then nylon mesh, and then the soil sample. The nylon mesh allows periodic removal and stirring of the soil sample to improve germination of dormant seeds, but without unwanted contamination by the non-experimental subsoil.

Although clayey soil samples require better water drainage, sandy soil samples need better water retention. This can be affected by lining the trays with vermiculate or peat, again separated from the experimental sandy soil with nylon mesh.

Clean vermiculate or peat also can be mixed with the sampled soil to increase water-holding capacity. In glasshouses, cold frames, and nurseries, shade cloth can be draped over the trays to retard evaporation and promote germination and emergence unencumbered by temporary water shortages. Trays also can be enveloped in clear plastic bags to maintain soil moisture; but this never should be done in sunlight, only in growth chambers where temperatures can be controlled precisely.

Most non-dormant seeds germinate quickly in the trays described above. Often 70 percent of the seedlings that will emerge eventually do so during the first two weeks of incubation, but this depends upon dormancy levels. All seedlings should be counted and removed as soon as most can be identified. The few seedlings that cannot be identified easily may be transplanted into pots for later identification.

After no further emergence occurs, it is common to mix the soil and begin another cycle of germination and emergence. The number of cycles varies among researchers. Roberts (1981) suggested continuing the process for two years. In our experience with summer annual species, most seeds germinate during the first cycle, about 10 percent of that number in the second cycle, and very few seeds germinate during a third or fourth cycle. There is no clear correspondence of species that germinate in one cycle but not another. Therefore, after two or three cycles we commonly stratify samples (4 C) for four or more weeks followed by alternating temperatures before returning them to the greenhouse. The purpose of this is to break dormancy in seeds that might have gone into secondary dormancy during the previous germination cycles. The intention is to mimic spring conditions that cause seeds of many summer annual species to be released from dormancy. Some workers place samples out of doors during winter for the same reason.

The time of year of sampling can influence how samples are handled. Samples taken right after seed rain might need a cold period or other stratification conditions prior to germination. Samples taken at the end of winter in temperate areas should be put in trays as soon as possible, as many seeds will have broken dormancy and are ready to germinate. Samples taken during mid-summer in temperate areas - after most weeds have emerged and before new seed rain - represent the persistent seed bank (Baskin and Baskin, 2000) which consists of mostly dormant seeds. These samples will probably require stratification and/or alternating temperatures to break dormancy and encourage germination. In the tropics, samples taken at the end of a growing season should contain freshly matured seeds that might need a dry period to break dormancy.

Some authors recommend sieving to reduce sample volume. Gravel and organic material larger than the largest anticipated seeds can easily be eliminated. If wet sieving is used, the mesh size must be very small if the smallest seeds are to be retained. For the silt loam soils we have worked with in Ohio, wet sieving results in a muddy plug of soil and organic material that is difficult to handle; therefore, this step was avoided. Nevertheless, some workers consider this step important for bulk reduction and enhancing germination of some species.

There have been few attempts to validate the germination method because it is very difficult to do. One could add a known number of viable seeds of a given species to a volume of soil to verify that the correct number of seedlings will emerge from the sample. However, field samples contain seeds of various age and dormancy status, and the difficulty in the procedure is obtaining seedlings from all such seeds, not from readily germinable seeds. Alternatively, one can screen seeds from the soil following 'exhaustive' germination in an effort to find viable seeds that did not respond to the procedure. Any attempt to calibrate the method would have to include all species of interest. One of the values of the germination method is the ability to obtain a comprehensive assessment of species, including many that occur relatively infrequently. Thus, calibration would be impractical for such analyses.

Many workers prefer germination methods over separation because of the many limitations of the latter. Germination methods are only less laborious to a degree. Several months generally are required to obtain data, making this method impractical for prediction of potential weed populations within a growing season.

Specialized knowledge is needed to accurately identify seedlings. There is the inevitable problem of viable but dormant seeds that do not germinate during the period of the germination test in spite of efforts to provide appropriate environmental conditions for breaking dormancy and germination. Some workers have used seed separation techniques after the germination method to attempt to separate these seeds. One of the main advantages of this method over counting is its utility for detecting a wide range of species and thus for community analysis. In samples of long-term replicated experiments including tillage and crop rotation treatments, we have detected germinable seeds of 20-30 species using this method, including several species that we did not expect to find, given the composition of the aboveground community. Some of the species were represented by only a few individuals in a single plot, and it is unlikely that these would have been detected and identified properly from seed counts.

SEED BURIAL STUDIES

Some of the most valuable information regarding seed bank behaviour has arisen through experiments in which weed seeds were buried purposefully. There are three basic approaches to these types of studies, which are named: 1) inverted bottle; 2) seed packet; and 3) seeded core methods.

Inverted bottle

This approach was pioneered by Beal and others at the turn of the preceding century. The method has not been used recently. As the name implies, seeds are place in sand-filled bottles, the bottles are inverted, and then buried in soil. Bottles are retrieved at intervals, sometimes 20-year intervals, and examined for germination and viability. Because the bottles are inverted, the seeds are not exposed to the same level of hydration, drying, and rehydration as seeds in field soils. Longevity appears abnormally long for many species examined in these experiments.

Seed packet

This common approach allows seeds to be exposed to near-natural conditions after being confined to bags typically constructed with decay-resistant nylon mesh. Bags often are buried at various depths and retrieved for analysis at various times after burial. In these experiments, seed longevity typically is much less than that observed through inverted bottle experiments. If seed packets are retrieved at sufficiently short intervals, loss of viability can be ascribed to germination (through observation of seedlings or seedling remains) or simple seed death. These studies have aided greatly the recent understanding that seed longevity under natural conditions often is less than five years.

Seeded cores

This technique involves exhuming a soil core and replacing it with soil devoid of seeds except for those purposefully added. These studies seemingly are more natural in that the added seeds are exposed to similar microclimate, microbial, and microfaunal conditions as seed packets, but also the macrofauna that would be excluded by small-mesh nylon bags. The detraction of this method is that seeded cores must be retrieved precisely so as not to include the natural seed bank inadvertently, and then processed in a manner identical to standard soil cores for seed bank studies.

VERTICAL AND HORIZONTAL SEED MOVEMENT/DISTRIBUTION

A number of seed bank studies in recent years documented and modelled seed movement. Typically, these studies involved tillage-induced movement of seeds, or coloured synthetic beads that mimic seeds. Initially, the focus of these studies was vertical movement of seeds caused by ploughs, chisels, disk, and no-till drills, with the understanding that burial depth is primarily a function of tillage implement. Deep burial was thought to be associated with a number of potentially important demographic processes, such as fatal germination and microclimate-imposed dormancy.

More recently, vertical movement of weed seeds caused by repeated tillage operations using any of a variety of tillage equipment has been studied and modelled (Cousens and Moss; 1990, Mead et al.1998; Staricka et al. 1990). These studies all point to a consistent trend; namely, similar implements bury seeds in similar proportions at similar depths regardless of soil type and location of the experiment (Forcella et al. 1994). This suggests a very satisfying universality of tillage-induced seed burial models.

Additional studies also examined horizontal movement of weed seeds as a result of tillage machinery. Although movement induced by tillage equipment can be appreciable, such horizontal displacement is very little in comparison to that caused by combine harvesters.

HOW TO SAMPLE ABOVEGROUND VEGETATION?

Many textbooks in Plant Ecology list protocols for sampling aboveground vegetation, and we will not attempt to re-examine this topic. However, with respect to relating seed banks to aboveground vegetation, some of our experiences may be helpful.

Timing of plant counts is important in terms of associating results with seed bank densities. Again, plant samples need to be taken at a logical time that follows, not precedes, sampling for seed banks. Depending upon goals of the research, counts may need to be made at various times of the crop cycle; e.g. (a) immediately before crop sowing; (b) 4 weeks after sowing; (c) at maximum crop leaf area index; (d) at crop harvest, and sometimes (e) after crop harvest too. The proportion of the total plant population that emerges prior to each count can vary substantially from site to site and year to year, depending upon microclimate.

The weeds present four weeks after crop sowing usually represent the most important proportion of the total weed population, at least from the standpoint of in-crop weed control.

The density represented by this proportion, however, may not correlate necessarily with seed bank density. In this case, researchers are advised to attempt a correlation between seed bank densities and weed densities at times a + b, a + b + c, b + c, and so forth. Only after these types of assessments have been made can researchers conclude that relationships exist or do not exist between seed bank densities and aboveground vegetation.

Quadrat sizes for plant counts are usually 50-5 000 times larger in surface area than soil cores. Consequently, a close correlation between seed and plant densities should not be expected (Cardina and Sparrow, 1996). For this reason and others, rank correlation may be more appropriate than regression for relating seed banks to aboveground vegetation.

Simple minimum-variance tests can determine the number of quadrats to be used quite quickly. In any event, multiple quadrats (e.g. 10 per plot, each 0.1 m2) are much preferred to a single quadrat (e.g. 1 m2) per plot.

Arrangement of quadrats probably is not too important provided that the placement of quadrats spans the length and width of the plot or field. Some authors place quadrats over or adjacent to the point where soil cores were taken. Although logical, this may have little practical effect on the results given the extreme aggregation of many seed banks.

CONCLUSIONS

Recent widespread interest in weed seed banks is reflected in the abundant research reported on this topic at the Third International Weed Science Congress (Anonymous, 2001), and symposia specifically devoted to seed banks sponsored by the Association of Applied Biologists in 1998 and 2003 (Reading, United Kingdom). With such enthusiasm for this topic, some guidance regarding sampling protocols and techniques may be useful, especially for younger scientists just beginning their research, or even older scientists with newfound interests in seed banks.

No single protocol or technique will have universal appeal, but there are a number of fine points of which seed bank researchers should be aware, and these are discussed in the report. These points are involved with sampling adequacy, sampling patterns, sampling times, seed viability testing, seed separation, seed and seedling identification, and aboveground vegetation sampling. If the guidelines provided in this report do not improve the results of future seed bank studies, we hope they at least will help alleviate some of the tedious work involved with this type of research.

PARAMETERS FOR WEED-CROP COMPETITION

In most developing countries, agriculture employs more than three-quarters of the labour force and provides a major source of GDP with a share of 35-40 percent (Maskey, 1997). A crucial policy issue is how to raise the income of resource-poor farmers without despoiling the natural resource base. A holistic approach based on Integrated Pest Management (IPM), as well as on sound economic principles, would provide a useful framework to protect both resources and farmers' income.

Since its adoption, IPM (Integrated Pest Management) and its component Integrated Weed Management (IWM), has become the basis for all FAO plant protection activities because it contributes directly towards the achievement of sustainable agriculture in developing countries.

It is worth noting that IWM presents some important differences in relation to other IPM sectors:

1. The weed flora usually includes several species that contemporarily infest the same plot/field so, in practice, it is normally necessary to estimate the overall loss caused by the whole set of species instead of the effect of a single species. However, situations do exist in which the infestation is monospecific, or can be considered as such in relation to the competitive results (e.g. wild oat infestations in wheat, barnyardgrass in rice, gramineae in dicotyledon crops already treated in pre- or post-emergence with herbicides for controlling broad-leaved weeds).

2. Weeds have a strong periodicity, with plants co-existing at different stages of development; each species and each development stage within a species has a different impact and different sensitivity to the control measures, especially chemical weed control.
3. Herbicides do not generally control a single species but more than one, each with a different level of efficacy; although herbicides with a highly specific action spectrum exist. It is therefore not possible merely to extend to weed control the approach used for some time in the entomology sector, where each insect is controlled with specific products.

To implement an IWM strategy successfully, weed management should match the specific problems in a field and therefore some basic knowledge on weed and crop ecology and biology is needed to correctly predict the impact of a weed infestation on crop yield. Within this context, weed-crop growth characteristics and the dynamics of weed emergence are important aspects (Akobundu, 1998; Forcella, 1998). Many farmers in the developing world are unaware of several aspects of weed interference and the best time for weed removal (Akobundu, 1998; Labrada, 1996 and 1998), although there are exceptions. Ellis-Jones et al. (1993) found widespread recognition in Zimbabwe of the importance of early weeding for both weed suppression and improvement of rainfall infiltration.

Weed germination patterns generally result in cohorts of seedlings emerging over an extended period of time and are heavily influenced by weather conditions, soil type and cropping system (Vleeshouwers, 1997). The initial emergence time differs from year to year and varies according to the species ecological requirements (mainly temperature and soil moisture content - Forcella et al. 1997).

It is also well established, and has been experimentally quantified for several crops and types of weed infestation (Zimdahl, 1988; Berti et al. 1996), that the relative time of crop-weed emergence and the time of weed removal strongly ifluence crop production.

On small-scale farms in developing countries more than 50 percent of labour time is devoted to weeding, and is mainly done by the women and children in the farmer's family. In traditional farming systems knowledge of the so-called 'critical period' of competition would enable farmers to make the most efficient use of limited labour resources. In conditions of medium-high weed pressure, the critical period is approximately centred on the first one-third of the crop growing cycle.

For example, several major arable crops of temperate climates (e.g. maize, soybean, sunflower) take 100-140 days after emergence (DAE) to mature and the critical period is often between 25-40 DAE (Zimdahl, 1988; Doll, 1994). Of course, the critical period varies with the relative competitivity of the crop and the weeds: the lower is the crop competitivity (and/or the higher the weed flora competitivity) the longer the period the crop must be kept weed-free to prevent significant yield losses.

COMPETITION AND CROP YIELD

The presence of weeds in a crop leads to an increased number of plants within a certain area. Given that the crop density is already set at a level that optimises yield for that cultivar in that environment, the presence of weeds will lead to a reduction in the average yield of the crop.

In an infested field it is possible to identify different components of the overall competitive effect:

- Intraspecific competition between plants of the cultivated species;
- Interspecific competition between plants of the cultivated species and weed species;
- Interspecific competition between plants of the different weed species;
- Intraspecific competition between plants of the same weed species.

PREDICTING YIELD LOSS CAUSED BY A SINGLE WEED SPECIES

Assuming that crop density does not vary significantly, i.e. that the effect of intraspecific competition between plants of the cultivated species can be considered as constant, a simplified situation can be analysed: the effect of a monospecific infestation on crop yield. The yield loss caused by the weeds can be predicted with an acceptable level of precision through the use of mathematical models.

These can be empirical or mechanistic. Empirical models are based on mathematical empirical relationships between some important independent variables (e.g. weed density or cover, growth rates, time of weed emergence in relation to that of the crop) and a dependent variable (usually crop yield) the variations of which should be interpreted and predicted. Mechanistic models are instead based on the growth processes (i.e. light interception, photosynthesis, dry matter partitioning between different plant parts of the competing plants (crop and weeds)); they are obviously much more complex than the former and require a good knowledge of the mechanisms and interactions involved in the crop-weed system and a large amount of data or information as inputs. These models have a limited direct practical use, but are good study and research tools from which much simpler empirical models can be derived.

Many empirical models have been developed to describe, and possibly predict, the effect of weeds in crops, many of which are based on the relationship between crop yield loss and weed density (Cousens, 1985b). In an attempt to solve the problems linked to predicting yield loss in relation to weed time of emergence, alternative models have been proposed based on the relative leaf area of the weeds and crop.

Prediction of yield loss on the basis of weed density

The model most widely used to describe the yield loss depending on the weed density is based on the rectangular hyperbola model (Cousens, 1985a):

$$Y_L = \frac{iD}{1 + \frac{iD}{a}}$$

where YL is the relative yield loss, D is the weed density, i is a parameter that represents the initial slope of the curve and a represents the maximum yield loss found with a very high weed density.

Figure: Rectangular hyperbola that links the relative yield loss to the density of a weed species.

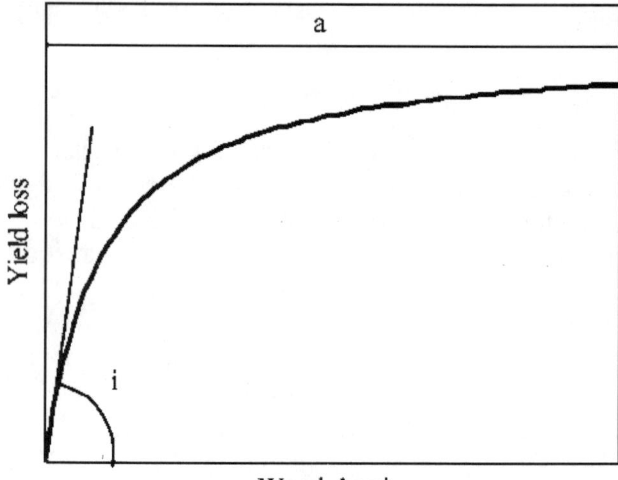

Both parameters also vary with the crop-weed association as well as with crop density, time of emergence of the weed and crop and soil fertility. The values of i and a can therefore be used to compare various crop-weed associations in additive competition experiments (i.e. when the weed is added to a crop sown at a fixed density).

Equation (1) provides the relative yield loss, that cannot be measured directly, but only calculated starting from the yields observed with and without weeds.

However, even the measurements taken in weed-free control plots can be affected by experimental error, so equation (1) can be adapted to include YWF among the equation parameters:

$$Y = Y_{WF}\left(1 - \frac{iD}{1 + \frac{iD}{a}}\right)$$

the third parameter YWF represents the yield of the weed-free control, which can thus be estimated using all the observed data and not just those from the control plots.

Equations previous assume that the weeds emerge contemporarily with the crop, but in practice this does not happen. Cousens et al. (1987) therefore introduced a modification to the above-described model to take into account contemporarily the weed density and relative time of emergence of the crop and weeds:

$$Y_L = \frac{iD}{e^{c \cdot te} + \frac{iD}{a}}$$

where YL, D, i and a have the same meaning as in (1), te is the relative time of emergence of the crop and weeds and c is the regression parameter that expresses the variation of the competitiveness of the weeds depending on the delay with which they emerge.

Model (3) is an improvement on (1) as it allows the experimental data to be better described when the emergence is not contemporary with that of the crop, but with very gradual emergences the determination of te becomes difficult and time-consuming and the estimate of the parameters arguable. Moreover, the estimate of parameters i and a, being dependent also on that of parameter te, generally show a higher variability over the years and with locations.

This drawback can, at least partly, be obviated using the temperature sum instead of days as a 'biological scale'.

Weed density is the variable most commonly used to explain the variations in crop yield loss. This has clear advantages, such as: a) simplicity of its control in experiments; b) determination in the field is relatively easy and fast. It also shows some disadvantages: a) in the field it is difficult to control the time of emergence of the weeds in relation to that of the crop, particularly when emergence occurs in flushes; b) the models based on it have weak eco-physiological bases.

For these reasons some authors have tried to identify another variable that would at the same time be easy to measure, take into account weed competitiveness determined by different times of emergence of crop and weeds, and be eco-physiologically sound.

Prediction of yield loss based on leaf area or relative weed cover

Kropff and Spitters (1991) proposed a relationship based on the relative leaf area (Lw) defined as the ratio between the leaf area index (LAI) of the weed and the total LAI of the crop plus weeds:

$$L_W = \frac{LAI_W}{LAI_c + LAI_W}$$

where LAIw and LAIc are the leaf area index of the weed and crop, respectively. Lw can vary from 0 (absence of weeds) to 1 (leaf cover of the weed alone). The authors, starting with the processes that regulate crop growth and the results of many simulations done with an eco-physiological competition model, demonstrate that the relative leaf area of the weeds compared to that of the crop at the moment of "row closure" could be a crucial measurement of the competitive process and very well correlated with the final yield loss.

The relationship between relative yield loss YL and Lw is expressed by:

$$Y_L = \frac{q\,Lw}{1+(q-1)Lw}$$

where q is an index of competitivity typical of a given weed in a crop, called relative damage coefficient.

Kropff et al. (1995) modified equation (5) by inserting a parameter m that represents the maximum damage caused by the weeds (asymptote of the hyperbole):

$$Y_L = \frac{q\,Lw}{1+\left(\frac{q}{m}-1\right)Lw}$$

Considering Lw instead of density takes into account, at least indirectly, that the damage caused by the weeds depends on the relative development of crop and weeds and therefore also their relative time of emergence. A certain value of Lw can be given by a few early-emerging plants or by many late-emerging ones.

The major problem of this approach is the difficulty in measuring Lw quickly and reliably. In fact, the precise measurement of leaf area is possible with destructive sampling of the vegetation (weeds and crop), but this method is neither rapid nor economical. An alternative is a visual survey, with the operator estimating the leaf area ratios between crop and weeds. With adequate training it is probably possible to obtain good results, but a level of subjectivity in the evaluations of Lw cannot be eliminated.

In place of Lw, the competitivity of an infestation can be evaluated on the basis of the partitioning of the upper layer of plant cover between crop and weeds (Relative Cover, RC). Observing the canopy vertically from above, the upper part of the plant cover will be formed by leaves of the crop and weeds: the ratio between the ground area covered by leaves of weeds and the total cover (i.e. weeds plus crop) represents the RC.

With W and C being the areas covered by weeds and crop, respectively, RC is given by:

$$RC = \frac{W}{W + C}$$

The higher this value is, the greater the share of solar radiation intercepted by the weeds will be, and therefore the competition caused by them will be more intense. This method assumes that: 1) interference for light is a measure of interference by all mechanisms: the leaf canopy may serve as an 'integrator' of the combined effects of competition for light, water and nutrients, and possibly also allelopathic effects, since these all reduce height, shoot weight and therefore leaf area and radiation interception; 2) the competitive effect of weeds that are shorter than the crop at canopy closure is negligible; in other words, only the plants that are able to overgrow or, at least, reach a height similar to the crop can successfully compete. The use of this variable was proposed with crops of medium-low height (e.g. peas, soybean): under these conditions, the use of the RC was shown to be valid. The applicability of this method still has to be evaluated with more widely spaced crops and on taller ones, such as maize and sunflower.

The ratio between RC and crop yield loss is similar to (6):

$$Y_L = \frac{q\,RC}{1 + \left(\frac{q}{m} - 1\right) RC}$$

Both Lw and RC proved to be good yield loss predictors at canopy closure stage, which is far too late for any control treatment. The early estimation of these yield loss predictors involves estimating their evolution from the time of assessment to canopy closure. This represents a further source of variability that has to be determined by means of appropriate experiments.

The main advantage of RC is the easier measurement. While Lw is based on LAI ratios and therefore requires all the leaf areas to be determined, RC requires the measurement of the ratios of cover between crop and weeds by observing the canopy vertically from above. This can either be done by means of a subjective visual estimate or, with greater precision, by measurements starting from photographs taken from 2-3 m above the crop. The possibility of using optical equipment that automatically take this measurement can also be foreseen.

It should be stressed that the basic data used for the determination of Lw and RC do not give any indication on weed flora composition, information that is fundamental for a correct choice of type of control (e.g. herbicide).

DAMAGE CAUSED BY MIXED INFESTATIONS

Information on single crop-weed associations, although interesting, is of limited practical interest because infestations normally include various species and the control measures, in particular herbicides, have a well-defined spectrum of efficacy.

The choices between control options (if and how to treat) must therefore be made, taking into account the damage that would be caused in the absence of control as well as the damage caused by the weeds that survive a given treatment. Both these evaluations require the competitive effect of a mixed infestation to be estimated.

In the case of the ratios between density and yield loss, the commonly used methods are based on the transformation of the observed densities into values that can be considered additive. It is obvious that two different weeds, even at the same density, will usually cause different yield losses: it is not therefore possible to directly sum up the observed density values to estimate the competitive effect of the infestation as a whole.

There are three main approaches: the first, proposed by Wilkerson et al. (1991) involves calculating the total competitive load (TCL). This method is used in the HERB program for the choice of post-emergence control options in maize and soybean. The second approach is based on the Density equivalent concept (Deq) and is the basis of a decision support system (GESTINF) adapted to Italian conditions and currently at the experimental stage at farm level. The third refers to relative cover and is still at the experimental stage.

Total competitive load method

The competitive ability of the various weed species is assigned through competition experiments by using an indexing method, i.e. assigning an arbitrary value K to the most competitive species and ranking the others according their relative competitivity with the reference species. Indexing is based on linear relationships between crop yield or biomass and weed density, which, for low density values, represent a good approximation of the hyperbolic relationship that exists between these two variables.

For low values of weed density (D), the crop yield can be given by:

$Y = a + b\,D$

The ratio b/a represents an index of the competitive ability of the considered species. A competitive index (CI) can therefore be defined for the ith species, obtained from the following:

$$CI_i = \frac{b_i/a_i}{b_r/a_r} K$$

where b_r and a_r are the regression parameters of the reference species.

K is a scale factor that can assume any value; the authors have chosen a value of 10. In this case, the weeds are ranked following a decimal scale with CI=10 for the reference species. CIs are the basic data used for the evaluation of the yield loss in a real field situation. Then, multiplying the density observed of the ith species by its CI, the competitive load (CL) for this species is obtained:

$CL_i = CI_i\,D_i$

The sum of these values for the various species present represents the TCL of that particular infestation:

$$TCL = \sum CL_i$$

Crop yield loss can be estimated on the basis of the TCL of the infestation. According to Wilkerson et al. (1991) the yield loss for low infestation levels can be calculated with a linear relationship. With the increase of the CL, the weeds start to interfere with one another as well as with the crop and the competitive effect caused by each single plant decreases. Under these conditions the crop yield loss follows a hyperbolic trend. In the case of soybean, Wilkerson et al. (1991) fixed the passing from the linear relationship to the hyperbolic one at a TCL of 50. It is worth mentioning that this model was calibrated for the central-southern United States, where soybean was grown at wide interrow spacing.

The complete expression of the yield loss as a function of TCL will therefore be:

$$Y_L = \begin{cases} 0.5\,TCL & \text{for } TCL \le 50 \\ 25 + \dfrac{55(TCL - 50)}{TCL + 60} & \text{for } TCL > 50 \end{cases}$$

Density equivalent method

The Deq of a given weed species is defined as the density of a reference species that determines a yield loss equal to that caused by the studied species at the measured density.

The crop yield loss in competition with the reference species is:

$$Y_L = \frac{i_{ref}\, D}{1 + \frac{i_{ref}\, D}{a_{ref}}}$$

While for the ith species present you have:

$$Y_L = \frac{i_i\, D}{1 + \frac{i_i\, D}{a_i}}$$

From the above definition, the Deq of the ith species (Deqi) is the value of the density of the reference species that makes the two preceding equations equal:

$$\frac{i_{ref} * Deq}{1 + \frac{i_{ref} * Deq}{a_{ref}}} = \frac{i_i * D_i}{1 + \frac{i_i * D_i}{a_i}}$$

A series of algebraic steps leads to:

$$Deq = \frac{i_i * D_i}{i_{ref} + i_{ref}\, i_i\, D_i \left(\frac{1}{a_i} - \frac{1}{a_{ref}} \right)}$$

The equation can be simplified choosing a hypothetical species as reference with the i and a parameters both equal to 1. This assumption gives:

$$Deq = \frac{i_i\, D_i}{1 + i_i\, D_i \left(\frac{1}{a_i} - 1 \right)}$$

Adding up the Deq of the various species present gives a total Density equivalent (Deqt):

$$\mathrm{Deq} = \sum_i \mathrm{Deq}$$

The crop yield loss will then be obtained from:

$$Y_L = \frac{\mathrm{Deq}_t}{1 + \mathrm{Deq}_t}$$

Knowing the indexes i and a of each weed species it is possible to calculate the damage that can be caused by any combination of these weeds.

Relative leaf area and relative weed cover methods.

If the competitivity of the infestation is evaluated using the Lw instead of density, the equation that expresses the effect of a mixed population becomes simpler. The formula that is applied for a single species can in fact be developed in an additive way, giving the following:

$$Y_L = \frac{\sum_i q_i \, \mathrm{Lw}_i}{1 + \sum_i \left(\frac{q_i}{m_i} - 1 \right) \mathrm{Lw}_i}$$

where qi, mi and Lwi indicate the relative damage coefficient, the asymptotic yield loss and observed value of Lw for the ith species, respectively. Where a maximum asymptotic value of yield loss is not considered, the value of mi is equal to 1 for all the species.

When using RC instead of Lw as a descriptive variable for weed-crop competition, the extension to a multi-species situation is straightforward. RC per se measures the ratio between the horizontal projections of crop and weed leaves, so integrating the effects of both leaf area and leaf posture of different weed species. This implies that RC is intrinsically a multi-species descriptor. Measurements taken in the field can therefore be directly used for estimating yield loss caused by a mixed weed infestation.

RELATIONSHIPS BETWEEN CROP YIELD LOSS AND WEED TIME OF EMERGENCE AND REMOVAL

Given the importance of early growth, the trend of the relationships between yield loss and time of emergence and removal of weeds can easily be understood. The competitive effect of a given density of weeds emerging with the crop depends strongly on the length of the period they remain in the field (i.e. the time of weed removal). The relationship between the duration of competition and crop yield reduction is approximately sigmoidal: weeds competing for a short period have little effect on crop yield; allowing the weeds to compete for a longer time, the yield reduction increases, until a plateau is reached corresponding to the yield loss caused by weeds competing over the entire growing cycle. Crops like maize and soybean show a relatively long initial period when the damage caused by weeds is relatively low, while most horticultural crops are more sensitive.

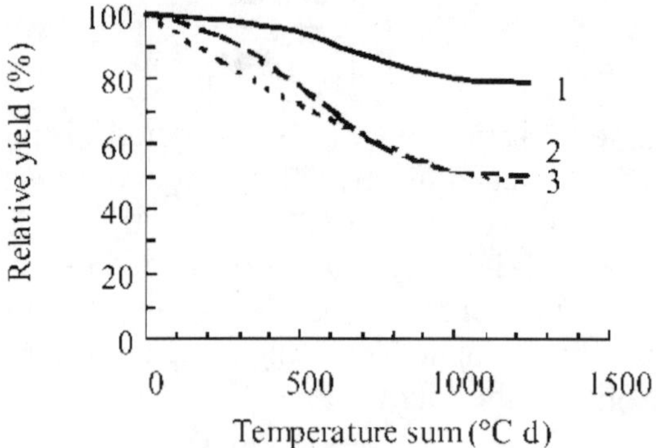

Fig. A: Crop relative yield, as a percentage of the weed-free test, as a function of duration of competition (i.e. time of weed removal) for a weed density of 10 plants m-2. 1 = maize in competition with Abutilon theophrasti (Sattin et al. 1992); 2 = soybean in competition with Amaranthus cruentus (Berti et al. 1990); 3 = onions in competition with Helianthus annuus(Dunan et al. 1995).

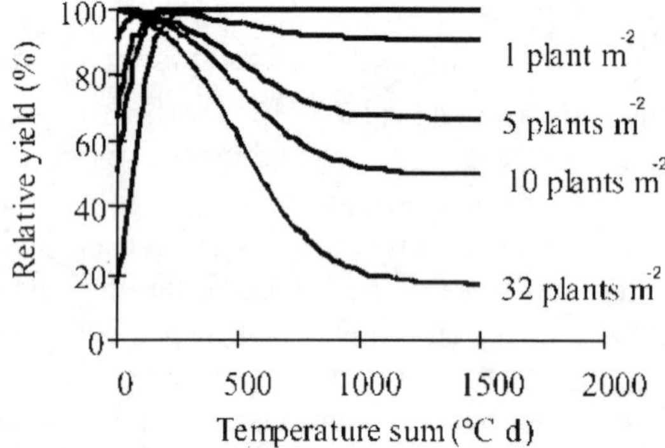

Fig. B: Soybean relative yield, as a percentage of the weed-free test, in relation to time of emergence, time of removal and density of Amaranthus cruentus (Berti et al. 1990).

The relationship between the time of emergence of weeds and crop yield loss mirrors the curves. Of course these relationships are influenced by weed density: for a specific weed time of emergence or removal, the higher the density the lower is the relative yield.

Extending the above concepts to a mixed-weed infestation that can be found in any cropped field, the yield can be expressed as a function of maximum yield of the crop kept weed-free, weed-competitive load and time of emergence and removal of the weeds. The first two factors are clearly site-specific, while contrasting data have been reported for the variability of the weed-free period (WFP) and duration of tolerated competition (DTC) curves for different years and/or locations. The experiments designed to obtain this sort of information are tedious and very expensive. An important question is therefore how variable are these data in space and time. Using three data sets showing the effects of mixed weed infestations in maize, soybean and durum wheat and by means of the Deq approach, Sattin et al. (1996) analysed the variability of DTC and WFP curves over reasonably homogeneous areas in temperate environments.

Despite the differences in weed flora among the experiments within each data set, the pattern of these relationships appeared to depend more on crop characteristics than on the composition of the weed infestation. If these results are confirmed for other areas and crops, for a given area the prediction of yield loss caused by a mixed-weed infestation with any time of emergence and/or removal will require the knowledge of the weed-free yield, the relationship weed competitive load-yield loss and only one set of parameters (i.e. only one large experiment) linked to the effects of the time of weed emergence and removal.

Information on weed competitive load-yield loss relationships already exists for several sites and crops and, if unavailable, it is much less tedious to obtain compared to the determination of the relationships between DTC, WFP, weed density and crop yield loss.

CRITICAL PERIOD AND OPTIMUM TIME FOR POST-EMERGENCE APPLICATION

The critical period has been defined as the period during which weeds must be controlled to prevent yield losses. Since the concept of critical period was introduced, it has been used to determine the period when control operations should be carried out to minimise yield losses for many crops (Zimdahl, 1988). Historically, critical periods have been calculated by mean separations (hereafter referred to as the classical approach) in experiments that evaluated the impact of time of weed emergence and time of removal on crop yields.

Using the classical approach, it is possible to identify a period within which no statistically detectable yield losses occur. It has also been concluded that for most field crops it is unnecessary to control weeds in the first few weeks after crop and weed emergence (Zimdahl, 1988).

Several problems inherent to the classical approach have been pointed out and the use of regression analysis (hereafter referred to as the functional approach) was suggested as a better alternative. In changing from the classical to functional approach, the existence of a period when weeds do not cause any yield reduction became doubtful because of the continuous relationship between yield loss and time of weed emergence and removal. To avoid this problem, fixed yield loss thresholds were used to define critical periods. Within this framework it was concluded that early weed control is unnecessary (Hall et al. 1992). This conclusion was not the result of an accurate evaluation of when to start to control weeds, but of the way critical periods were calculated. The functional approach fails to recognise that since there is a continuous relationship between crop yield and time of weed removal, controlling weeds pre-plant or pre-emergence or post-emergence cannot be compared without considering yield losses that occur between planting and post-emergence control.

The establishment of a fixed yield loss threshold indirectly considers the economic aspect in the calculation of critical periods. Within this framework, Dunan et al. (1995) developed an economic approach to calculate critical period.

They defined the economic critical period as the time interval when the marginal income of weed control is higher than the cost of control, and its limits are called early and late economic period thresholds.

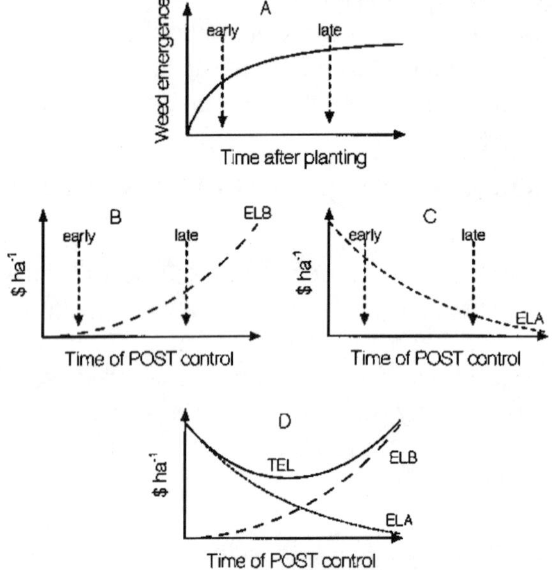

Fig. Theoretical representation of the concept of optimum time of application for post-emergence treatments.

- Hypothetical pattern of weed emergence as a function of time after crop planting.
- Economic loss as a result of weeds emerging before the post-emergence treatment (ELB).
- Economic loss as a result of weeds emerging after the post-emergence treatment (ELA); and
- Total economic loss (TEL) as a function of time of control. Early and late represent two different times of post-emergence weed control

and it is assumed that the efficacy of post-emergence treatment is 100 percent.

The period when weed control can be carried out starts before sowing with a pre-plant weed control tactic and continues until the phenological stage of the crop, which precludes any further weed control. Within this time lapse the three approaches (classical, functional and economic) all define a period when weeds should be controlled.

However, identification of a period alone does not provide information on how many or when control practices should be performed. Considering the time dependence of the weed competitive effect, treatments carried out within this period cannot be regarded as having the same net margin (i.e. difference between the value of the crop with control minus the cost of treatment and the value of the crop without control) and therefore an optimum time for one or more weed control tactics exists which provides the highest net margin.

Weed emergence can begin from the moment of final seedbed preparation and then continue while environmental conditions are favourable for the germination process.

As previously shown, the time of weed emergence greatly affects weed competitiveness, with the first emerging weeds being far more competitive than the late emerging ones. Most post-emergence treatments have low or nil residual activity. Because of this they can control the weed population present at the moment of application, but have little or no effect on subsequent germinations. The economic result of post-emergence treatments will vary according to the time of application and is related to the efficacy of the treatment, crop characteristics (WFP and DTC curves) and weed germination pattern.

As a first approximation, crop yield loss is given by the sum of the yield loss caused by weeds emerging before the treatment (competing with the crop until the treatment) and that caused by weeds emerging after the treatment (competing from emergence until harvest). These two yield losses follow opposite trends.

With an early treatment, the loss caused by the weeds emerged before the spraying is low because the duration of competition is short. On the other hand, a consistent number of weeds can germinate after the treatment and, remaining until harvest, can produce an important yield loss. With a late treatment, damage caused by weeds emerging after the treatment is reduced, but there is a marked increase in the damage as a result of the weeds emerging before the treatment.

The sum of these two losses gives a total economic loss curve characterised by a minimum which identifies the optimum time of treatment (Berti et al. 1996).

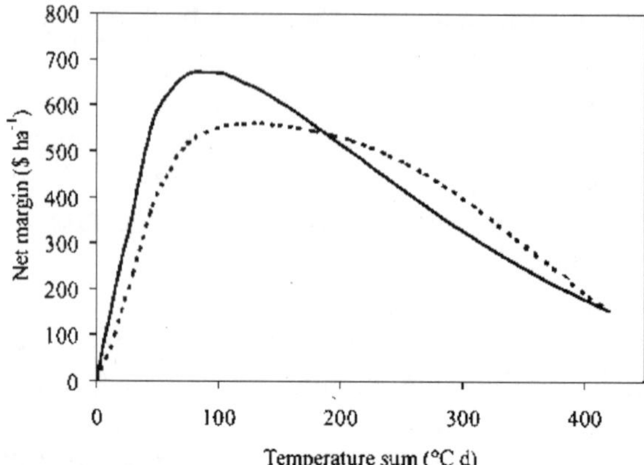

Fig. Relationships between time of application of a post-emergence treatment and net margin for maize and soybean. Maize: weed-free yield = 10 t ha-1; cost of post treatment = 45$ ha-1; grain price = 155$ t-1. Soybean: weed-free yield = 4 t ha-1; cost of post treatment = 45$ ha-1; grain price = 250$ t-1. For both crops a weed infestation capable of causing a yield loss of 50 percent if untreated was considered.

An example of these calculations for maize in the Po valley - Italy and for soybean in southern Ontario - Canada (Sattin et al. 1996). The curves were calculated using the Time Density Equivalent approach (TDE - Berti et al. 1996) and the DTC and WFP relationships calculated by Sattin et al. (1996) were used. With the TDE approach, the weed population is divided in daily cohorts depending on their time of emergence. For each cohort the TDE is equal to the density of a cohort of weeds emerging with the crop and competing until harvest, giving the same yield loss as the considered cohort. TDEs can be considered additive and their sum is a Total Time Density Equivalent (TTDE), which gives a good estimate of the competitivity of the weed population depending on the emergence pattern and, considering herbicide efficacy, on the time of treatment. The relationship between time of weed control and net margin depends on the pattern of weed emergence and on the competitiveness of the crop.

THE PROBLEM OF WEED-PRESSURE ASSESSMENT

The approaches presented above would allow to decide if, how and when to treat. This obviously requires some knowledge on the quantitative and qualitative composition of the weed flora. Depending on the approach used, different measurements should be taken (i.e. weed density, weed and crop LAI, relative cover). This phase of scouting is crucial: it is clear that the higher the level of precision the better it is, being the basis for subsequent calculations and for the selection of the weed control option to be adopted; on the other hand, scouting requires time and represents a cost, decreasing the net margin of weed control. Furthermore, weeds usually have a patchy distribution in fields and this requires the sampling techniques to account for this. In the case of weed counts, Berti et al. (1992) proposed a relatively easy sampling method. Scouting is done by throwing a metal frame (25 × 30 cm) at random: the plants present within the rectangle are counted and classified by species. On the basis of previous studies (Berti et al. 1992), it has been verified that 20-30 throws are enough to guarantee a good estimate of weed density, assuming that the infestation is relatively homogeneous; if not sub-samples should be taken. The time required for this type of scouting is around 20 minutes per hectare when it is done by well-trained personnel, but can be more than doubled with less skilled people (Berti et al. 2002). The time requirement and the need for specialised personnel can become a major problem in the application of Decision Support Systems (DSS) in the field, especially in developing countries.

Other approaches have been proposed based on visual estimates of weed pressure. In the United States, Harvey and Wagner (1994) used a visual estimate of weed pressure for estimating crop yield loss. Weed pressure is defined as the visual estimate of the percentage which weeds contribute to the total volume of both crop and weeds in a plot, with a value of 0 indicating the complete absence of weeds and 100 indicating the complete absence of crop plants. A common problem with this type of approach is the high variability of ratings assigned by different personnel for the same level of infestation, leading to an uncertainty of yield loss estimates, particularly with low levels of infestation, which represents, on the other hand, the situation where the use of a DSS is more useful. A further constraint is that the visual estimation of weed pressure is easily feasible when both crop and weeds have already reached an advanced stage of development, while it is very difficult to give a correct rating when plants are at the seedling stage. This limits the usefulness of this approach for the selection of present-year weed-control measures, while it might be used to predict future crop yield losses in the same field in order to assist in making sound weed-management decisions for the following year (Harvey and Wagner, 1994).

Therefore, at the present time the use of DSS for selecting weed-control measures faces a paradox: They are especially useful to less skilled farmers, because they can greatly improve their management techniques, while more experienced people can normally select the 'right' control measure based on their own experience. On the other hand, to be done rapidly and reliably, thus not determining an excessive increase of control costs, a proper assessment of weed competitive pressure requires a relatively high level of knowledge.

CONCLUSIONS

The use of models relating weed density or other independent variables expressing weed competitivity can greatly improve the weed control selection procedure, with a great impact on both yield and economics. However, applications of DSS derived from these competition models are, so far, relatively limited, mainly because the use of DSS requires a relatively high level of knowledge to be really effective.

The wide literature on this subject shows that the beginning of the growing cycle is crucial in determining the intensity and outcome of the subsequent weed-crop competition. It also appears that in hotter climates where the crop growing cycle is often shorter, as is found in many developing countries, the importance of early weed-crop competition is accentuated. The relative time of weed and crop emergence appears to be crucial and can overwhelm many other factors: a few days' difference in emergence can create an unrecoverable gap between plants, and therefore any agronomic practice which delays weed emergence and favours a satisfactory crop establishment plays an important role. There are indications that the variability (in space and time) of yield loss caused by mixed weed infestations can be relatively low. Therefore, if these results are confirmed for different areas, for various crops and for a reasonably homogeneous area, only a few low-tech experiments, although labour-intensive, are required to predict yield loss in relation to WFP, DTC and weed density.

This type of information could give useful indications and provide a framework particularly in situations where the technological level is low (i.e. where hand-weeding is prevalent), and then where simple rules directing weed control are more likely to be adopted instead of more complicated computer-based technologies. It would therefore be very useful to have, at least for the most important agricultural areas and crops, some basic data sets related to weed-crop competitivity, timing of weed competition and weather conditions.

GUIDELINES FOR WEED-RISK ASSESSMENT IN DEVELOPING COUNTRIES

The movement of trade goods and aid of one sort or another throughout the world is essential for the well-being of all peoples. Not all these goods and gifts are benign, however, and some come with unwelcome surprises. Invasive species affect agricultural and other systems, and their impacts are second only to habitat destruction in terms of loss of biodiversity. These concerns have generated growing international interest in weed-risk assessment systems to prevent the introduction of new pests and to prioritise existing pests for control. Weed-risk assessment is a new discipline, and the first international symposium on the topic was held only recently in Australia (Groves et al. 2001). This country, along with New Zealand, is at the forefront of developing and implementing strong quarantine protocols. Both countries are relatively isolated from the rest of the world, agriculture is important to their economies, and their citizens value natural landscapes and their ancient indigenous biodiversity.

This report introduces the topic of weed-risk assessment and provides guidelines for countries wishing to strengthen their own quarantine protocols and to use scarce resources efficiently for prioritising existing pests for control. Fortunately, this task is becoming easier because the Internet allows rapid exchange of information and access to detailed databases on pests, e.g. the global compendium of weeds.

INTERNATIONAL FRAMEWORK OF WEED-RISK ASSESSMENT

The actions taken to exclude a plant species from a country because of its weed potential must be consistent with the international standards regulating the movement of trade goods. These obligations are defined under the Agreement on the Application of Sanitary and Phytosanitary Measures (SPS agreement) of the World Trade Organisation (WTO 1994), and the International. Plant Protection Convention (IPPC) (1997 revised edition) deposited with the United Nations Food and Agricultural Organization (FAO, 1996). These two international agreements, whilst allowing countries to specify requirements for the entry of plant material, describe the obligations of countries so that import requirements are not unjustified trade barriers.

A further international convention involving weeds concerns the need to conserve biodiversity. Convention on Biological Diversity states that: "Each Contracting Party shall, as far as possible and appropriate, prevent the introduction, control or eradicate those alien species which threaten ecosystems, habitats, or species." Not all countries are signatories to this convention.

A quarantine pest is defined by the IPPC, as a pest of potential economic importance in an area endangered thereby and not yet present there, or present but not widely distributed and being officially controlled (FAO, 2001a). It is accepted for the purposes of this report, that 'economic importance' includes actual or potential effects on the economy of ecosystems and their component species, and that the IPPC definition of a pest is sufficiently broad to include weeds covering the full range of ecosystems, including those covered by the Convention of Biological Diversity (CBD 2001). In fact, international meetings have recently been held to foster collaboration between the IPPC and the CBD.

Pest Risk Analysis (PRA) is a three-stage, process of evaluating biological or other scientific and economic evidence to determine whether a pest should be regulated and the strength of any phytosanitary measures to be taken against it (FAO, 2001b).

These stages are:

Stage 1. Initiating the process by identifying a pest that may qualify as a quarantine pest, and/or pathways that may allow introduction or spread of a quarantine pest that should be considered for risk analysis in a defined PRA area.

Stage 2. Assessing the pest risk by determining which pest(s) are quarantine pests, and characterising the likelihood of entry, establishment, spread, and economic importance.

Stage 3. Managing the pest risk identified in Stage 2 by developing, evaluating, comparing, and selecting options for dealing with the risk.

The initial steps are to determine the pathway(s), that is, any means that allows the entry or spread of a pest, and correctly identify the pest. The identification of high-risk pathways is an important part of an overall weed-risk assessment process, but this report deals only with the individual pests.

The criteria used to determine the presence or absence of the potential quarantine pest in the area. Area is defined as an officially defined country, part of a country, or all or parts of several countries (FAO, 2001a). If the species is absent, and has potential economic importance, it can be considered a quarantine pest. If it is already present in an area, then it can be legitimately considered a quarantine pest and evaluated further if it is of limited distribution or under officialcontrol. Official is defined as established, authorised, or performed by a national plant protection agency, and control is defined as suppression, containment, or eradication of a pest population (FAO, 2001a). A pest capable of further spread, that is, expansion of the geographical distribution of a pest within an area, that is not controlled, would require to be put under control to justify quarantine pest status. Species that are controlled but are at the absolute limits of their potential distribution cannot spread further and so cannot be declared quarantine pests either. In reality, most exotic species in most countries have potential for further spread.

Once the quarantine pest status has been confirmed, the next step is to assess the economic (including the environment) importance of the species. This may be high for a pest.

Weed-risk assessment is concerned primarily with the first two stages of the pest risk assessment involving pest categorization, that is, the process for determining whether a pest has, or has not, the characteristics of a quarantine pest or those of a regulated non-quarantine pest (FAO, 2001a). The minimum requirement of any weed-risk assessment system is that it satisfies the international agreements. To do this it must be built on explicit assumptions and must use scientific data.

Weed-risk assessment systems designed for use only within a single sovereign state, and which do not have the potential to limit trade, need not comply with international agreements. But to be effective they must be based on similar sound principles.

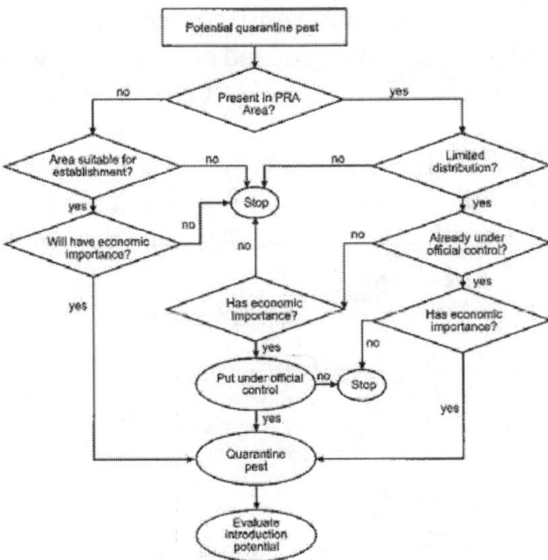

Fig. Pest risk analysis (from FAO, 1996).

PLANT INVASIONS AS A PROCESS

Plant species must cross a series of barriers to reach a new area and spread within it. Initially, these are physical barriers on an intercontinental and/or intracontinental scale. Species that have not crossed these barriers may nevertheless be classified as quarantine pests, primarily on the basis of their pest history elsewhere.

Once they have reached the new area, they must overcome a range of abiotic and biotic barriers before establishment. Human activities are important in assisting species to cross these barriers. Species arriving in small numbers by accident have a relatively low chance of establishment. In contrast, those species spread widely as seed contaminants, raised in large numbers within a protected environment for horticulture, or planted out in agricultural or natural environments, e.g. crops or erosion control, have a greater chance of establishment. Once a species is growing in cultivation in a new country it may spasmodically appear in the wild beyond the initial plantings. If it was introduced occasionally as a contaminant of crops it may appear on associated land. The term for such sightings is casual alien or casual exotic. To become a naturalized alien or fully naturalized alien or exotic depending on the definition being followed, a species must then develop self-maintaining populations in the wild. These loci are the points from which it may spread within the area.

Different species characteristics and life-cycle stages may be important at different barriers. For example, colourful flowers may be the selection criterion for transcontinental transportation in the first place. Rapid reproduction by seed and/or vegetative offspring (e.g. bulbs and tubers) may then assist its spread once it has been introduced. Its persistence through periods of unfavourable climate may depend on long-lived seed banks. Crossing any of these series of barriers is reversible. A species may be extirpated locally or even driven to extinction within an area, if, for example, there are severe climatic fluctuations or new predators and diseases are introduced. The process of arrival and extinction by natural means or control may be repeated over many years until the species finally becomes fully naturalised.

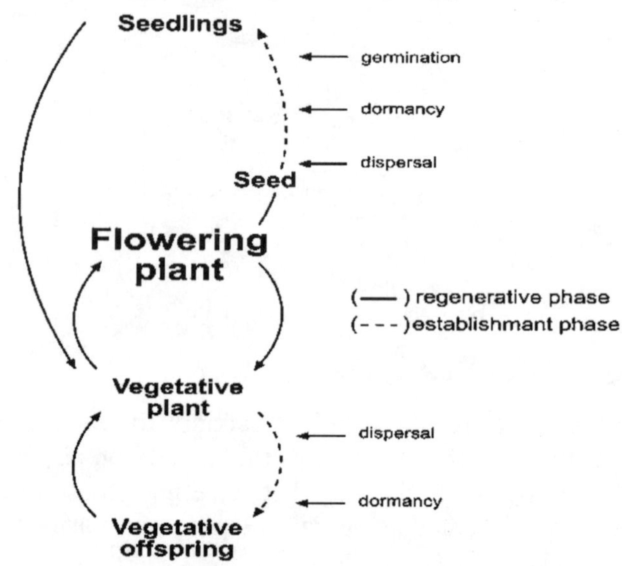

Fig. Life cycle of perennial plants producing both seeds and vegetative organs

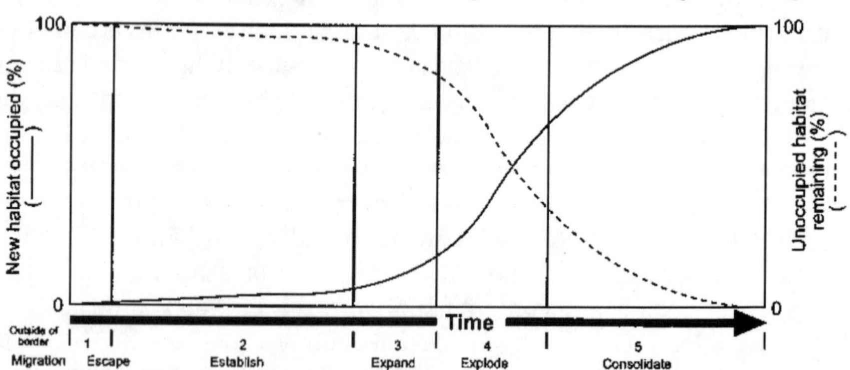

Fig. Conceptual phases in the invasion of a weed through time, and the way these relate to the percentage of occupied and unoccupied land.

The spread of a pest may follow a number of patterns in time and space, depending on such factors as its means of dispersal, life cycle, and so on. Many follow a simplified 'S' shaped pattern that can be illustrated graphically as the proportion of all potential habitat occupied by the pest at any point in time. The essential features are a long tail at the beginning of a species spread as it crosses the first series of barriers, a steep rise as it breaks through these barriers and finds suitable habitats, and then a flattening off as these habitats are saturated.

As the pest spreads, the proportion of the uninfested habitat declines at a rate defined by a 'reverse S'. The process of spread may be continuous, but points are still recognisable (usually only with hindsight) where the rate of change alters markedly from the preceding period. For management purposes the 'S' shape can be idealised as stages based on the extent and rate of spread. This concept can be applied at any geographic scale, from a field to a continent.

CONCEPTUAL PHASES IN THE INVASION OF A WEED

Migration phase

The species must first reach the border of the area. Once it has arrived it may, or may not, enter, depending on a variety of factors.

Where there are efficient quarantine protocols and risk management procedures it will be detected and, it is hoped, eliminated, if a quarantine pest.

Escape phase

Once inside the area it may escape only occasionally, or finally become fully naturalised. The locations of these naturalisation points are likely to be associated with the pathway of introduction, e.g. in fields planted with contaminated corn, or adjacent to erosion-control plantings. They have been referred to as 'sentinel sites.'

Establishment phase

During this phase, the plant is able to reproduce in the new environment, and population numbers slowly begin to build up. Virtually all potential habitat is still uninfected.

Expansion phase

Eventually, the number of sites occupied expands beyond the initial loci. Expansion is fastest where there are multiple loci. The causes of this expansion differ among species and are not well documented. Factors are diverse, including particularly favourable growing seasons, the arrival of new pollinators or dispersers, the species becoming adapted to its new environment by the formation of new genotypes. New habitats may be created, e.g. by changes in land use.

Some local areas of habitat are noticeably infested, but most potential habitat is un-infested. It is often only at this stage that the plant begins to be perceived as a pest.

Explosion phase

The period where the pest expands rapidly and often where it begins to attract official concern. Many potential habitats are infested during this phase.

Entrenchment phase

The pest slowly spreads to the last remaining habitats over its full range within the area. This does not mean that it occurs on all suitable land at any one time, but that it has a high chance of occurring there.

Further spread can occur only if more suitable habitat is created, e.g. by fire. Importantly, the pest may be present only in a dormant stage of its life cycle.

These potential changes in the spread of a pest have implications for weed-risk assessment imperatives:

- The most cost-effective means of avoiding pest impacts is to prevent their introduction or establishment in an area. Failing that, the greatest return for expenditure of money and effort comes from controlling a pest before it has spread.
- Once it has established and begun to spread, the ongoing effort required to eliminate it increases dramatically.
- During the earliest spread phases, when the required funds to extirpate a pest are low, these may be effectively obtained as an adjunct to other pest control programmes. Once the pests begin to spread rapidly, the effort required to obtain the funds may be orders of magnitude greater. To rephrase this, at one extreme, only the person with the spray gun (or spade or slasher) needs to be persuaded to act, while at the other, it may require the approval of government.
- There are no recorded successful weed extermination attempts worldwide where the pest covered more than a few hectares. Species with persistent seed or other regenerative life features that require repeated visits to the site(s) are particularly intransigent.
- The total accumulative costs over time are the effort required obtaining the funds, the money spent on actual control, plus the impacts on the economy and environment. These accumulated costs become progressively greater with time if control attempts are delayed, as illustrated by the differences in the three shaded curves.
- Effective weed-risk assessment systems must be appropriate to: whether or not a pest has established and spread; the pest's biology and ecology; the values being threatened; the extent to which it has have or has not established in an area; and the technologies and resources available.

Figure: Relative combined monetary and environmental costs of undertaking an eradication programme (A), together with those of initiating ongoing control programmes at an early (B) and late stage (C) of the invasion. Arrows indicate programme starting points. The differences in area beneath the curves (B-A, C-A, C-B) represent the benefit of control action at the earlier stage.

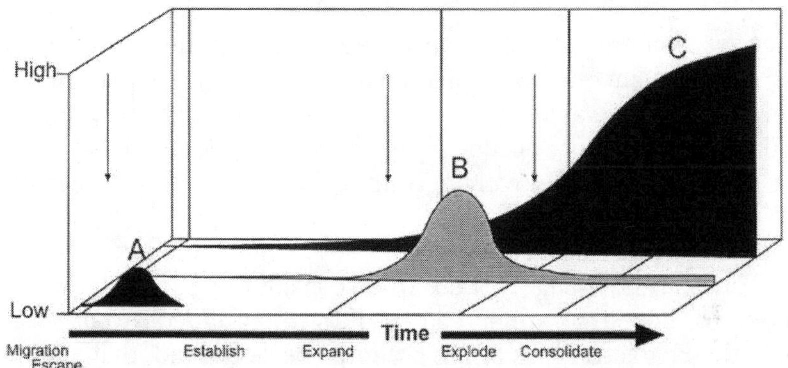

Three important factors change as a pest crosses the various barriers:

a) The location and amount of information. Before a species has been introduced to an area, all information needed to undertake pest categorisation will be derived from experience of the species outside the new area. In most cases, this will mean obtaining information from its country of origin and elsewhere. The quality and amount of information will depend largely on whether the species has a history as a pest, or perhaps as a crop or ornamental. Once a species has been introduced, much information can be obtained, particularly about its growth and reproductive biology. Only as it spreads, however, will its environmental tolerances and impacts within the new area be revealed.

b) The certainty of a correct assessment. Resulting from this increasing information as the pest spreads, the reliability of a weed-risk assessment increases, and conversely, the chance of the assessment being incorrect, decreases.

c) Identification of those affected. The ecosystems on which a pest might impact cannot be reliably predicted before it has begun to spread. Neither then, can the individuals or interest groups directly affected by the potential impacts be identified. As a pest spreads through its potential habitats and range, those affected become increasingly identifiable. The corollary of this situation is that those who are likely to benefit from the management of the pest become increasingly identifiable. It means that at one extreme, all those within the area potentially benefit from the detection of a quarantine pest that has not reached there. At the other extreme, only those whose land the pest occupies benefit directly from the local control of a widespread pest.

It is ironic that weed-risk assessments have the greatest chance of being wrong when they are most effective in preventing accumulative impacts and costs of control, and in potentially benefiting the widest range of interest groups. That some risk is acceptable, however, is recognised by the third stage of the pest risk assessment process concerned with managing the risk.

LIMITATIONS OF WEED-RISK ASSESSMENT

Risk assessment systems concerned with both quarantine pests and established pests aim only to predict the potential harmful effects of a species. The weighing of these aspects against any potential beneficial outcomes, e.g. production of a new crop, or food in a shipment of contaminated grain, is an entirely separate exercise involving value judgements. It is not a component of weed-risk assessment as such.

Two important issues are faced in selecting a weed-risk assessment system. These are particularly acute when considering species new to an area as opposed to those that are spreading. First, it is difficult to predict a pest from only the characteristics of the potential pest. Second, that amongst any group of organisms, those attaining pest status do so at a very low rate.

Many studies have attempted to identify the characteristics of pests, as distinct from benign species. Early studies attempted to identify an 'ideal weed'. More recent studies have concentrated on groups of similar plants within a country, a continent, or throughout the world. A few variables are associated with weediness that is broadly applicable over whole groups of plants, e.g. herbaceous agricultural weeds, or plant families, e.g. pines. These rarely have predictive value when extended to broader groupings, as from agricultural systems to the natural environment, from pines to non-pines amongst the conifers. The consensus appears to be, that no traits are universally important for all species in all habitats. The characteristics of the receiving environment are equally important.

The importance of any particular plant trait in determining the success or failure of invasion becomes discernible only after the species has either established or known to have failed in a new habitat. As the species' fate becomes apparent through time, the reliability of the prediction will therefore increase.

Even then, there may be no suit of endogenous plant characters readily obtainable from the literature that reliably predicts potential weediness. However, the chance of a plant species establishing in a new area is related to the pressure of its propagules on that area. This may be defined as the rate of individual whole plants, or vegetative or sexual reproductive parts, that are dispersed into an area over a given time period. Propagule pressure can operate at any special scale. Examples of low propagule pressure are the infrequent arrival from year to year of an occasional wind-blown seed from a distant source, the occasional seed of an unwanted species in a seed lot, or the infrequent planting of a timber species that produces few seeds. Examples of high propagule pressure are the frequent arrival on a regular basis of wind-blown seeds from a distant source, abundant seeds of an unwanted species in a widely planted seed lot, or abundant viable and readily dispersible seed from widespread plantings of timber trees.

The proportion of imported species that become pests has been calculated as ranging from about 0.01:100 for British angiosperms, 1.3: 100 for grasses into tropical Australia, and 12.0: 100 for pines in New Zealand. However, this wide diversity of ratios, and the effects of time lags between establishment and species acquiring recognised pest status changing the ratios, means the search for any constant ratio is futile. As a consequence of the low proportion of pests amongst a random selection of related organisms, any system designed to detect pests is likely to be wrong as often as it is right. The results of a false positive assessment (excluding a species when in fact it would not become a pest) could have long-term economic consequences for a country, e.g. in the case of a potential new crop. In contrast, a false negative assessment (accepting a species that becomes a pest) could result in serious economic damage.

It is equally important to realise that introductions supposedly for beneficial purposes also have a very low success rate. For example, hundreds of grasses and herbs were introduced into Australia and New Zealand, yet the agriculture of these countries is based on only a handful of species. Similarly, the exotic forest plantations of New Zealand and southern South America are dominated by a single species, Pinus radiata. Furthermore, some estimate of economic benefit can be predicted for agricultural/forestry systems assuming a certain level of uptake, or deleterious effects assuming a certain level of spread as a pest. In contrast, there is no adequate ecological theory to underpin predictions of the future impact of potential environmental weeds. There is great difficulty therefore, in predicting both the positive and negative effects of new introductions.

Despite the apparent theoretical impasse in the prediction of pest status, the task must be undertaken because of the consequences of not detecting potential new pests at the border, of benign species becoming pests, or of the spread of existing pests. Weed screening systems have been devised for woody plants in general (Reichard and Hamilton, 1997), groups of woody plants (Tucker and Richardson, 1995), and water plants. A few systems are used by national risk assessment authorities for use as quarantine tools, i.e. the United States Department of Agriculture (Lehtonen, 2001) and the Australian Quarantine Inspection Service (AQIS). A common failing of these systems, is that they do not calculate the probability of the realisation of a predicted impact. Such predictions can be made only with large sample sizes of all individual cases in a class. They would require large databases of known plant histories. In New Zealand for example, where the total number of exotic species is known, the probability of a species naturalising - the first step to having an impact - has been calculated for all families and genera. These data can be incorporated in weed-risk assessments.

Weed-risk assessment systems have limitations, but their widespread use will encourage the international recognition of weed-risk assessment as a discipline.

Risk assessment systems operating at the border of an area to detect quarantine pests not yet established there, and those assessing established pests, have potential fundamental differences related to managing the risk. Managing newly-detected quarantine pests may mean simply prohibiting the entry of the species. There may be costs in doing so, related, for example, to loss of profits from the sale of a shipment of seed containing the pest, or a potential new agricultural crop foregone because of the potential pest status of the species. However, for any additional species undergoing weed-risk assessment there will be relatively little opportunity cost to the administrating authorities unless it involves monitoring a new pathway. In contrast, the outcome of a risk assessment identifying an entirely new pest within an area may require substantial management expenditure to extirpate the infestation. Unless additional funds are available to do this, they will need to be reallocated from elsewhere, usually from other pest-control efforts. Thus there is a need to determine the potential impacts relative to those of existing pests. Internal weed-risk assessment systems therefore havepriority of management and expenditure as an important component of their process.

PROCEDURES FOR QUARANTINE WEED-RISK ASSESSMENT

The Australian WRA system

The Weed-Risk Assessment (WRA) system was developed in Australia and is the most widely known and applied border weed-risk assessment system encompassing all plant groups. The central 'argument' is that if a species has had the opportunity to become a weed in another country, and it has done so, then it should be classed as a weed. Provided, that is, that the climate and environment are compatible with the new country (they are assumed to be so if there is no information). While this argument is essentially circular - it is a weed elsewhere, therefore it will become one here - a history of weediness elsewhere has reliably predicted weediness in several studies (Scott and Panetta, 1993; Reichard and Hamilton 1997; Williamson, 1998; Maillet and Lopez-Gacia, 2000). The WRA system was tested in Hawaii where it was found to be the most successful of those compared. Until such time as weed-risk theory can contribute greater precision to the practice of weed-risk assessment, the WRA system is a suitable tool for use as a quarantine tool in developing countries. This argument is further reinforced by the imperatives of these countries to protect productive lands of one sort or another from weeds likely to arrive from developed countries (or via their neighbours) in trade goods, e.g. grass seed for sowing, or as a result of aid projects, e.g. re-vegetation schemes fostering legume shrubs. Because the actual or potential weediness of such species is commonly recognised in the developed countries where these goods or schemes originate, the central argument of weed history will be a powerful one in identifying quarantine pests in the developing countries. Ironically, the reverse is increasingly not the case. For example, in the last 20 years, 70 plant species have naturalized in New Zealand that are not known in the wild in any other country outside their native range. These were introduced mostly for urban horticulture and they come mostly from developing countries where their potential weediness has not been realised, e.g. Cotoneaster spp. from China.

The WRA produces a score for weediness and converts this into an entry recommendation for a specified taxon. It also satisfies several other requirements of an acceptable biosecurity assessment system (Hazard, 1988; Panetta, 1993). It can be calibrated and validated against a large number of taxa already present in the recipient country. These should represent the full spectrum of taxa likely to be encountered as imports into that country. It has some success in discriminating between weeds and non-weeds, such that the majority of weeds are not accepted, non-weeds are not rejected, and the proportion of taxa requiring further evaluation is kept to a minimum. The system also identifies which major land use system the taxon is likely to invade. This aids an economic evaluation of its potential impacts. In this respect it appears to be more successful at identifying agricultural weeds than environmental weeds. The WRA attempts to separate economic plants from those unlikely to have any economic benefit in a new country. However, where a taxon may have significant economic benefits, economic value should be scored in a transparently separate exercise and balanced against weediness in appropriate risk assessment evaluations.

The WRA is not obligatorily computer based, but when operated on a computer it becomes interactive. This allows assessors to measure the influence of different attribute values on the scores generated. Finally, the system has proved to be cost effective to prospective importers and to border control authorities in Australia and New Zealand.

Permitted list approach

The process in which the WRA operates as described here is based on the concept of a Permitted List of plant species (or defined taxa). This system is used by quarantine authorities in Australia (Walton 2001) and New Zealand.

The underlying concept is that if a species, or any subspecific taxon, with the potential to be a pest in an area is not on a list of taxa permitted to be in that area, then it will be prohibited until it has undergone pest categorization. Many countries, for example the United States of America, have lists of quarantine species, but they do not have permitted lists.

They do not determine if every species new to a country should be on either list. One advantage of the involvement of a permitted list is that it automatically triggers a pest risk analysis in circumstances where there might otherwise not have been an analysis.

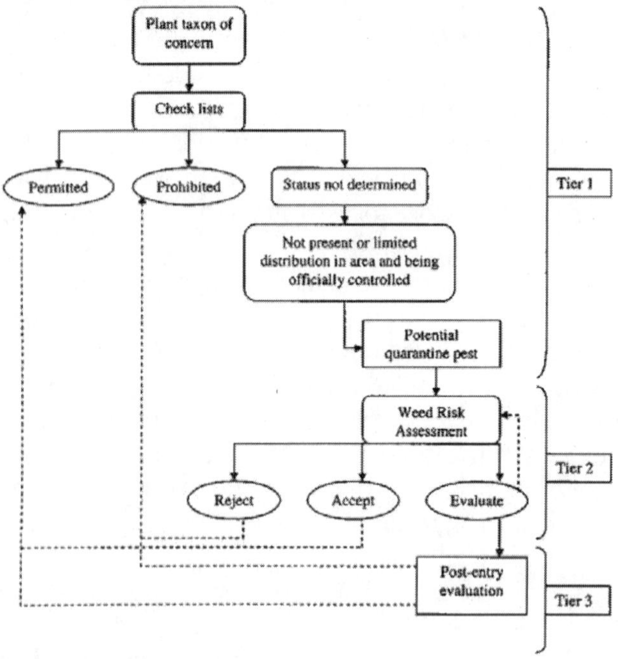

Fig. A flow chart for screening plant introductions incorporating the permitted list approach.

There are three stages to the prohibited list approach (Walton, 2001).

Tier 1. The first task is to identify a taxon correctly and determine whether it is listed as prohibited or permitted. This requires checking its species, genus, and family names, and whether there are synonyms. The next step is to check its presence in the country, either in cultivation only or in the wild. If a species is neither permitted nor prohibited, its spread within the area and whether or not it is under official control have to be determined. The kinds of information necessary to determine.

In some cases, the gathering of information on its status in the area may reveal both its presence, and sufficient evidence to justify an internal pest risk analysis. The outcome may result in the taxon being subjected to official control.

Those with limited spread would automatically go on the prohibited list. Data should be entered into appropriate databases at all stages.

Tier 2. Once a taxon has been classified as a potential quarantine pest, it undergoes a weed-risk assessment. This involves the WRA system, which recommends a species be rejected,accepted, or undergoes further evaluation. Rejected or accepted species are added to the prohibited and permitted lists respectively. Otherwise, further evaluation may be required if the importer wishes to proceed.

Tier 3. A classification of 'accept' or 'reject' cannot always be obtained from the second tier after the gathering of further information. It may then be appropriate to conduct field or glasshouse trials to further evaluate a species. These would need to be conducted in a secure environment from which the species could not escape via wind-blown seed, for example, or 'hide' in persistent seed banks. Whether the prospective importer considers this additional effort is warranted may depend on the potential gains from the plant species.

WRA operation

The WRA is based on the answers to forty-nine questions. These cover a range of weedy attributes in order to screen for taxa that are likely to become weeds of the environment and/or agriculture. The questions are divided into three sections producing identifiable scores that contribute to the total score.

PROCEDURES FOR INTERNAL WEED-RISK ASSESSMENT

Choice of a system

The objectives in characterizing potential quarantine pests are relatively straightforward because species are uniformly at their migration phase. In contrast, the objectives and information requirements of an internal weed-risk assessment system change as the species spreads.

Decisions are made at an increasingly local level. Politics and economics may enter increasingly into the analysis as beneficiaries of control become identifiable and competition for resources between sectors increases. These latter issues are not dealt with exhaustively here, but see Wainger and King (2001).

The selection of internal weed-risk assessment systems must therefore consider the spread stage(s) of all pests being compared, impacts on the systems they affect, the likely benefits (and beneficiaries) of control efforts, and the quality of the available information. These factors vary widely within countries and between countries. There are numerous internal weed-risk systems in use. Often several are in use simultaneously within one country, even at a national level.

Table. The main systems used primarily for internal wee -risk assessment and prioritizing.

Author(s)	Approach
Champion & Clayton, 2001	Scores for plants ecology, biology, and weediness of aquatic weeds
Esler *et al.* 1993	Sums scores for ability to succeed with a score for weediness
Hierbert, 1997	Weighs relative impact against ease of control and cost of delay
Randall 2000	Scores for invasiveness/ impacts/ potential distribution/invasion stage
Tucker & Richardson, 1995	Models attributes of species and matches them with environment
Timmins & Owen, 2001	Explicit weed-led approach cf. site-led. Considers value of area potentially impacted
Virtue *et al.* 2001	Multiplies scores for invasiveness, impacts, and distribution (current and potential)
Wainger & King, 2001	Relates likelihood of damage/defined functions of landscape/and the scale of threat to appropriate response

A sound and practicable system for application at a local level is that of Randall, (2000). More precise and ecologically defensible systems focus on specific biomes, such as the shrub lands of South Africa (Tucker and Richardson, 1995) or aquatic weeds. Aquatic systems are almost a class on their own and authorities should be circumspect about any new wetland species.

Overall, generalised systems are probably required first in most countries. Besides, the detailed understanding of the relationship between species attributes and the environment that make biome-specific systems effective are poorly understood for most biomes.

A single, widely applicable system cannot be recommended until the objectives of the internal weed-risk assessment system are determined. Countries establishing weed-risk assessment systems at the national level should ensure that the data collected are applicable to a range of spread stages and spatial scales of weed control. National resources for assessment made available by central government should be allocated where the long-term benefits will be greatest. This means first a border screening system, next a system for prioritizing species in the early establishment or expansion phases, and only then to species that have consolidated their spread. For this last group, the detailed system of Virtue et al. (2001) at a national level could be improved only with considerable effort, but would need to be adapted to the area concerned.

Factors to consider

This chapter describes the factors to be considered, the kind of information needed and likely to be available at different spread stages, and the process of determining what kind of weed-risk assessment system is required in specific sets of circumstances. Many of these considerations are relevant to a range of scales, from a single property to a whole country, and they are always limited by the total amount of resources for pest control within the particular area.

Weed history of congeners

Weed-risk assessment systems developed for the border, e.g. the WRA, usually consider the weediness of an assessed species' relatives as indicators of potential pest status. This factor has seldom been considered as a risk component of systems designed for species at the earliest spread stages. Exceptions are where this association is implied by the group of species being ranked, e.g. pines (Tucker and Richardson, 1995). The behaviour of species' relatives (e.g. family, genus) at several levels of taxonomic grouping may usefully be incorporated in internal weed-risk assessments. This applies particularly to species at their earliest invasion stages, where in the absence of much other information, it may contribute to a stated probability of a successful invasion.

Weediness is concentrated within certain genera in some families and widely dispersed among many genera in others. Whether these probability estimates can be made at the family level or sub-family level depends on the size of the plant family and genera. Many genera are too small to give statistically reliable ratios.

Weed-led and site-led control

There is a tendency to control only those familiar species that have traditionally been controlled. This inertia demands a prioritising system that reallocates resources away from individual species that have become uncontrollable at a defined scale, to those that are potentially controllable at the same scale. Once attempts to extirpate a species or reduce it to below a defined population density over the entirety of a defined area (species-led control) have failed, then it should be controlled only in specific high value places within the area (site-led control).

The concept was developed for conservation weeds in New Zealand (Williams 1997) and the application of this principle to crown owned conservation land is explained by Timmins and Owen (2001). It is relevant to a range of systems, including agricultural systems, and can be used in prioritising pests to be controlled at a national level.

Invasion stage

Some estimate of a species' stage of infestation, or its surrogate, is required to determine the practicality of control. A clue to spread rate may be indicated by resident time within the area, if this is known, compared with the present distribution. However, unless a plant species is already listed as an unwanted organism for a specific area, often only range expansion prompts the realisation of a new potential pest. By this stage, most newly recognised pests are well established and spreading. Where historical distributions are unknown, the simplest approach to the infestation stage that avoids the difficult interpretative question of spread rate, is to ask how well the species is established, i.e. the present number, size, and distribution of infestations. This also relates most closely to potential control of the species - those spreading rapidly will usually be well established with many loci, and will be more costly to control if widespread. These factors of range and expansion rate need to be considered within the context of the species' regeneration time. A species does not necessarily spread to the most favourable or potentially damaging habitats first. Consideration needs to be given to the more favourable habitats and/or more vulnerable land uses it might encounter as it spreads.

PREREQUISITES FOR PEST EXTIRPATION

Systems to determine whether the species is a candidate for weed-led or site-led control need to result in a 'yes' or 'no' outcome. Predictions of management outcomes may be more reliable than those concerning more complex ecosystem and economic interactions. The question, "Can it be killed?", is easier to answer than "Will it affect biodiversity?", or, "What economic impact will it have?" Even for well-established serious weeds, particularly in natural systems, the most meaningful trigger for their management may also be determined primarily by the cost and efficacy of control measures (Panetta and James, 1999).

Species extermination has seldom been achieved over areas of greater than a few hectares anywhere in the world. Irrespective of the area covered or the perceived impacts a pest may be having, there are critical questions for preventing, selecting and determining the level of management:

- Can all individuals of the species be located; and are practicable control measures available?
- Can all individuals be targeted within a defined time period dictated by the life cycle of the plant?
- Are the plant's responses to control known?
- Are resources to treat new plants, at least as fast as they appear, available to undertake follow up work?

Those corresponding to 'yes' answers to these questions are higher priority for weed-led control than those with 'no' answers. Information on several aspects of a species is necessary before being able to answer these, and other, questions.

Biological attributes

A cautionary note can be made here. A wide range of biological attributes have been used in attempts to characterise weediness and prioritise species for control. The WRA also considers these factors too. Most detailed biological attributes are only assumed to equate to invasiveness, even if this is taken to mean spread, as opposed to impacts. While some very general rules relating species attributes to invasiveness are emerging, these apply to only a few groups of plants in specific habitats. These often have particular disturbance regimes, including those determined by human activities. In many natural or semi-natural systems the relative importance of various dispersal modes is unknown. For example, until the relative number of potential wind-dispersed and fleshy-fruited woody plants potentially available to colonise lowland wooded vegetation in New Zealand are known, it is uncertain certain whether the fleshy-fruited syndrome per se has lead to the relative abundance of the latter group. Thus, attributes such as dispersal mode may be used with more integrity if used indirectly in determining management options, such as search frequency, rather than in attempting to predict invasion rates per se.

Ease of eradication

The intensity of weed control can be thought of as the product of the difficulty of killing an individual at the first event, including such factors as non-target effects, multiplied by the frequency of visits to re-treat the infestation. If a plant species has a history of weediness then it is likely to have certain identifiable attributes that make it so, e.g. persistent seed bank, and to have been the subject of control attempts elsewhere. These can help assess the difficulty of control in the new area. Where there is no history even of cultivation, or of weediness in its home region, ease of eradication must be inferred from attributes of the species or its congeners. These attributes could be classified in a variety of ways, but four seem to be critically important.

Time of detection

Detection of new infestations within one or two generations is important if the species is to be eradicated or contained within a small area. This means that the species has to be recognisable as a weed at an early stage. Species cryptic in the wild, such as a short grass or a vine with inconspicuous foliage, are likely to be confused with desirable species by the moderately informed observer. They are likely to spread before they are identified as weeds. They will be more difficult to control than conspicuous species.

Reproductive capacity

The amount of viable seed and vegetative reproduction may be critical components of invasion success. However, there is less certainty about the relative importance of these factors to invasion, or to ease of eradication. Species with persistent seed banks can be just as difficult to eradicate as those lacking seed banks but with vegetative reproduction. There is evidence that species with more than one reproductive system are more invasive, on average, than those with only one system. This is partly because the different strategies may enable the species to cross a wider range of barriers to invasion. As the population increases the barriers change. Species can therefore be ranked according to the number of reproductive strategies they have, without making assumptions about the relative importance of these strategies.

Dispersability

Potential for dispersal is obviously essential but the relative importance of different dispersal mechanisms should not be overstated in assessing weed-risk - most plants have a dispersal system of some sort. Wind-blown seeds are commonly blown long distances, and small seeds, more than large seeds, can be consumed and dispersed by wider-ranging animals. But seed size must always be considered within the context of the range of available dispersers and the potential dispersal mechanisms within the area. Passive dispersal by water, machinery, etc., and through contaminants in produce, may be more important than biological characteristics. Alternatively they may interact, e.g. small seeds are more likely to be carried by machinery than large seeds. In assessing risk from invasion, likely dispersal routes (waterways, farm tracks, randomly) also need to be considered, along with the suitability of the surrounding landscape to the species. In the early stages of invasion, an important contribution of dispersability to weediness is the ability to hide, as discussed above.

People

The attitudes of people towards plant species vary widely. While some species are considered a nuisance by everyone, others are useful to various sectors of the community. The outcomes of human activities involving these useful species, including recognised pests, can have an overriding influence on their spread. Attitudes to a species need to be considered, and as a rule, those species favoured for one reason or another will be the most difficult to extirpate.

For a species to qualify for a programme aimed at extirpating it from a defined area, the probability of re-invasions from outside sources should be nil or very low. This is often not possible for those species grown commercially that are also pests.

Here it may be possible to minimise the risk to land beyond the plantings by preventing the species regenerating within the defined area. This option may apply when a decision is made that the benefit emerging from a new species outweighs the risk, and pest risk management procedures are put in place while the species is grown commercially.

Climate matching

The utility of climate matching to weed-risk assessment changes as a pest spreads. At the earliest stages only the broadest match between source area (native and or adventive range) and potential range is required to consider the species a potential pest, because climate may or may not be the major barrier limiting spread. Many grasses originally from tropical Africa, for example, are now widespread in temperate regions. At the latter stages and on a local scale, climate matching is less important because the species has shown its potential to spread, and the ranking systems are required only to prioritise amongst known pests. Thus, climate matching between current and predicted range is most useful as a prioritising tool at the intermediate stages of spread, particularly when viewed on a country scale. Climate matching requires thorough distributional data within the areas being considered. On a country-by-country basis this requires comprehensive national databases. If data were collected on a regional basis, e.g. southern South America, perhaps under the umbrella of organisations such as COSAVE [The South Cone Plant Protection Committee (Comité de Sanidad Vegetal del Cono Sur - COSAVE)] it would have greater utility than on merely a national basis. Climate-matching is a specialist activity beyond the scope of this report, but the reader is referred to Kriticos and Randall (2001) for a summary of the applicability of several software packages to this topic in Austrialia.

Impact

Pests have economic, ecological, and or/social impacts, and the assessment method must define which of these it is attempting to assess. Reliable estimates of impact are possible only after the pest has begun to spread. Estimates of impact often involve a calculation of unit impact times a measure of the area covered. Several kinds of impact may be determined, or estimated, for one or more species, and incorporated into a scoring system (Virtue et al. 2001). Impacts may be determined on a very coarse scale and equated merely with presence, e.g. a species is present in 'x' number of land use systems in 'y' number of regions of a country. Much finer scales may be used, and extrapolated over the potential range of the species, e.g. a weed is sprayed at a cost of 'w' dollars on 'x' number of ha that would amount to 'z' dollars over its potential range.

Species impacts elsewhere may be applicable in the new area. In the absence of history, impact has to be estimated from the attributes of the species. These will differ with the land use likely to be affected. In agricultural systems, the impact of related pests may be relevant. For conservation land however, there is no universally applicable measure of impact. Parker et al. (1999) proposed parameters that might eventually be quantified as: I (overall impact)= R (range) × A (abundance) × E (impact per capita).

Species vary widely in the biomass at maturity that can be generated from a single propagule (seed) or ramett (piece of stem or root). An estimate of the biomass and extent of a species can contribute to a rudimentary estimate of impact. There may be evidence of its growth rate in terms of height and area covered. They are likely to range over tens of orders of magnitude, e.g. from a single grass plant 10 cm tall by 25 cm2 (0.002 m3), to a typical perennial herb, 1 m2 and 1 m tall (1 m3), to trees 10 m tall and with crowns 10 m diameter (1 000 m3). 'E' is likely to be related to the log of the volume of a single individual plant: 1, 10, 100, 1 000, 10 000. These data can be reduced to scores ranging from 1-5.

Biomass as a surrogate measure of impact is probably modified by the species' physical interaction with desirable vegetation. Information is generally available on whether the species co-occurs or replaces the desired vegetation. The long-term effects of weeds in either in the canopy or in a lower regenerative layer are mostly unknown. Intuition suggests replacing the vegetation canopy will, in the short term, displace more species, including invertebrates, than simply occupying a sub-canopy position. This generalization may not hold all species, e.g. for herbaceous ones increasing the effects of fire in natural systems. Similarly, impact is likely to be related to persistence at the site, whether this is via a single generation, or via successive generations.

DESIGNING A SCORING FORMAT

A scoring or ranking system for internal weed-risk assessments should embody all the principles of a quarantine assessment system, other than the requirement to meet international obligations (unless they were likely to impact on international trade). It should be designed to produce ranks, or other forms of classification.

These should be based on the premise that weed management options are a function of the magnitude of risk, that the greatest benefit is achieved by controlling populations at the earliest stages of invasion, and that the score will be modified by the position of the manager to reduce the risk. It should identify the invasion stage(s) targeted and the ecosystems potentially affected. It should be no more comprehensive than is necessary to utilise the available information. Because weed control technology and resources available to manage the risk can change, these should be considered as separate modules and incorporated into the decision-process.

Ranking systems in use differ in the information required to operate them, and also in the structure of their internal rules. The simplest systems give numerical ratings to a set of criteria that may or may not be divided into sections, and are then summed. The questions may have equal or unequal value. The individual scores may, or may not, be modified by the answers to others. The subtotals from one section may be modified by other subtotals (Owen, 1998; Randall, 2000). Aspects of the species sometimes appear twice, as in its innate ability to be a pest and the ease with which it can be controlled (Hierbert, 1997).

There may or may not be default scores where questions are not answered, and points may be deducted if answers to certain questions are negative. Others operate via hierarchical decision trees (Reichard and Hamilton, 1997). In a completely different approach, Tucker and Richardson (1995) used an expert system where a series of questions filtered out species of high or low risk before progressing to the next question.

An internal weed-risk assessment system should confirm, more or less, the existing ranking of weeds within an area for predefined spread stage, if this has been undertaken by experts, rather than produce a reordering of priority species. In other words, the outcomes from any new system must be intuitively sound if the system is to gain acceptance and be applied. This approach then captures all knowledge of the weeds of an area and formalises it within a system that is transparent, repeatable, and applicable to newly recognised species.

The development of these systems on spreadsheets allows the component scores to be adjusted and the effects on species rankings examined.

RESOURCES REQUIRED FOR WEED-RISK ASSESSMENT

A weed-risk assessment is primarily a 'book exercise' involving the collection of all available information about a potential pest, interpreting it, and making a decision. There are basically three kinds of information necessary, as well as access to the Internet for gathering it:

- The primary resource required to comply with the IPPC standards is a list of exotic plant species in the country, and whether they are under official control. To collect such information for both cultivated and wild species is a huge task and it has been achieved in very few countries. The first priority for most countries lacking this information is to restrict the list only to exotic plants under official control. Their names must be valid, and ideally there should be a voucher specimen. Once this is achieved, the task can be expanded to include wild exotic plants not under control. These tasks require the close cooperation of a range of national and regional organisations including herbaria, universities, libraries, and agriculture departments.
- Books are an important source of information on plants. Those undertaking weed-risk assessments should have access to all the relevant regional floras and the numerous books on weeds of the world or of regional weed lists. A basic collection of such books. The two most important are Holm et al. (1979) describing the world's

worst weeds, and Huxley (1999), which lists all the families, most of the genera, and many of the world's species.

- Online information sources are now critical to weed-risk assessment. They can be used for checking identification and plant names, determining weed history and noxious status elsewhere, and investigating control methods. E-mail is also a critical tool for communicating with numerous experts around the world who are familiar with these resources. In many cases they can also supply information on species not recorded in any form.
- Computer software is not critical for weed-risk assessment on a case-by-case basis but it is potentially a very important tool. It is necessary, however, if climate-matching tools are to be used. Apart from this purpose, the basic software requirements are database software and, if it was decided to use the Australian WRA system, a recent of version of EXCEL.

4

Ethnobotanical Investigation of Weed Biodiversity and Rice Production

BACKGROUND

The collection and consumption of 'wild' plant foods from agricultural and non-agricultural ecosystems has been documented in multiple cultural contexts, illustrating their use and importance among farming households throughout the world. The evidence to date suggests that gathering by farmers occurs in various environments, ranging from intensively farmed areas, to more subsistence oriented horticultural systems, and finally in more pristine areas such as forests. This is certainly the case of rice farmers in Asia. For example, Ogle et al. found that in the Mekong Delta of Vietnam 90% of women eat wild vegetables, uncovering a total of 94 species. Kosaka, in his research on flora from the paddy rice fields in Savannakhet, Laos, recorded 11 edible species from a total of 19 herbaceous useful plants, and 25 food trees out of 86 useful species. The documentation of 'wild' food plant gathering and consumption in mainland Southeast Asia is still growing, however the literature is scattered across numerous disciplines.

The research on which this paper is based was conducted in Kalasin Province, Northeast Thailand. Studies conducted in this region provide documentation that 'wild' food plants are a critical component in the subsistence system of farmers. This food resource is extremely important to the rural population comprised of rice farmers, given that the Northeast region is regarded as both Thailand's largest and poorest part of the country. This paper adds to this literature by providing the most comprehensive botanical inventory of these foods to date. Two botanical characteristics are described in this chapter: growth form and life cycle. Moreover, we present the growth location of the plants. Regarding cultural characteristics, this paper also identifies multiple uses of wild food plants.

Wild food plants in this chapter refer to non-domesticated plants. These plants exist on a continuum of people and plants interactions in regard to their degree of management. In this way, wild food plants include those from 'truly' wild to wild protected, cultivated and semi-domesticated plants that may be promoted, protected or tolerated in some way locally. Wild food plants can be cultivated, but not all cultivated plants are domesticated. For most species the transition from cultivation to domestication never happens. Human plant management does not necessarily move toward greater intensity and ultimately plant domestication. While some plants are moving towards domestication, other plants that used to be highly managed in the past could be only slightly tolerated and protected under contemporary circumstances. While we include in our definition 'introduced' and 'naturalized' plants, locally domesticated plants are excluded. We use the term 'local' because, since the nature of this study is ethnobotanical, we based our research on these plants that are classified as 'wild' by local people. This is why some food plants that are regarded as 'wild' in Kalasin, might be treated as domesticates in other areas.

THE RESEARCH SITE

The research for this paper was conducted in four villages in Kalasin Province, Northeast Thailand. The villages are fairly typical for the region. Kalasin is located at 152 m above sea level (asl) in the Korat Plateau, which geographically defines the Northeast region of the country. This Plateau, forming a shallow depression between 100 m and 200 m asl, is generally quite flat with scattered swamps and ponds (some seasonal) and low hills that rise to around 300 m asl.

Soils in this region are mostly heavily leached fine sandy loams, with poor drainage and high salinity. Furthermore they are usually low in phosphate, nitrogen and organic matter. Declining soil fertility is prevalent in the region. Nevertheless, the soils in lowland paddy fields are better than in the uplands because they receive nutrient in-flows eroded from the higher areas. The natural vegetation of this region is dry monsoon forest, primarily composed by dry dipterocarp forest, with Dipterocarpus tuberculatus Roxb., D. obtusifolius Teijsm. ex Miq.,Shorea obtusa Wall., S. siamensis Miq., Xylia xylocarpa (Roxb.) Taub., Irvingia malayana Oliv. ex A.M. Bennett, Cratoxylon formosum (Jack) Dyer. and Careya arborea Roxb. as dominant species.

Deforestation has been occurring at a high rate since the early 1950s with the extension of agricultural land due to commercialization of agriculture, as well as population growth.

In this way, the forest and wooded areas have decreased from 90% in the 1930s to less than 14% in 2004. The rate of deforestation was likely augmented significantly during the economic crisis at the end of the 1990s. At the same time, soil degradation in the agricultural areas has been increasing and consequently yields have declined.

The Northeast covers 170, 000 km2 and has more land dedicated to agriculture than the rest of the country (9.25 million hectares). Around 94% of the region's population live in rural areas, with the region possessing the highest number of farms in the Nation (2, 273, 000). Indeed, in Kalasin province 85.1% of the population depend on agriculture. The main crop is glutinous rice (also called sticky rice), which is important as the dietary staple and for income generation. Rice production corresponds to 70% of the arable land of the Northeast, but average rice yields are the lowest in the country (1.8 Mg ha-1). Within the traditional rain-fed paddy agricultural system, which is primarily transplanted rice, crops can be damaged by delayed rains when transplanting seedlings, or by droughts and floods. The annual monsoon provides 90% of the annual rainfall of the Northeast, averaging over 200 mm from May through October, which is essential for the cultivation of glutinous rice. From November to April, rainfall averages only about 20 mm per month in Kalasin.

THE RESEARCH POPULATION

The Northeast is referred to as Isaan and is also known for its distinct cultural characteristics. The people who inhabit the region, commonly referred to as Isaan people, are ethnically of Lao origin, constituting one of the largest minority populations in the country. Most Northeasterners speak a dialect of Lao mixed with some influences from Thai also known as Isaan. Isaan is written using the Thai script. Thai is learned formally in school and villagers are literate in Thai, except for the very elderly.

Kalasin Province has a population of about one million inhabitants and a density of 132.3 inhabitants/km2. Households on average have four family members in the rural areas, and 23.6% of them are female headed. Theravada Buddhism is the main religion in this province (99.5% of the population), as in the rest of the country. The population has attained on average 6.5 years of education. Regarding their work status, 51.7% are unpaid family workers and 35.8% are engaged in self-employment, usually in agriculture. There is a high rate of seasonal or full-time migration to major cities mainly as wage labourers who aim to send remittances to their families that stay in the rural areas. Off-farm employment accounts for two thirds of the total income of families in Northeast Thailand. There is customary inheritance of land through women and a pattern of matrilocal residence. This system facilitates women having a thorough knowledge of their social and physical environment.

GENERAL OVERVIEW OF WILD FOOD PLANTS IN NORTHEAST THAILAND

An important yet not widely available study at the national level established that wild food plants play an essential role in the diet in all the rural areas of Thailand. This is clearly reflected in the fact that more than 500 different edible natural products have been documented as being sold in the markets around the country.

Gathering mainly occurs in anthropogenic ecosystems, such as agricultural lands (including paddy fields), woody areas, (home) gardens, house areas and swamps. Agricultural lands and home gardens are traditionally owned by women. In Northeast Thailand, women are the main gatherers, selectors, transplanters and propagators of wild food plants.

In this region farmers have as their staple glutinous rice accompanied by a variety of wild foods derived from wild, semi-domesticated and domesticated plants, as well as frogs, paddy crabs, insects and fish. During the rainy season wild food can constitute as much as half of the total food consumed in the villages. Wild food plants are mainly consumed as fresh fruits or vegetables eaten raw or steamed, and in local "curries" or soups.

In fieldwork conducted in Northeast Thailand in 1990, Price documented 77 species gathered by farmers in a village in Kalasin Province. Somnasang, Rathakette and Rathanapanya listed 42 wild vegetables and 7 wild fruits in a paper published in the 1980s. Ten years later, Somnasang, Moreno-Black and Chusil recorded 66 wild food plants consumed in Northeast Thailand. Furthermore, Sapjareun, Kumkrang and Deewised published a book, in Thai, entitled "Local vegetables in Isaan" presenting a general description of a number of plants by species, their propagation, ecological importance and uses, as well as the local recipes.

THE BOTANICAL-DIETARY PARADOX

One of the important characteristics of wild plant foods among farming households is that the main collection locations are increasingly from the anthropogenic ecosystems such as agricultural areas rather than pristine ecosystems. Ogle and Grivetti in their study in Swaziland found that the most intensively cultivated area among their research sites exhibited the highest level of loss of edible species, but, at the same time, the most consumption of wild food plants.

They termed this phenomenon the "botanical-dietary paradox" and proposed that this occurs when people start to rely on eating the weeds of agriculture once a decline in forests occurs. Ultimately, the species that are considered local vegetables change. Price and Ogle further explain that time constraints are a major factor in the commencement of the botanical-dietary paradox in that as forests decrease and become more remote from the village, gathering from the forests becomes increasingly too time consuming, so farmers shift to gathering in areas closer to home and shift to eating many of the weeds of agriculture and other food plants in the agricultural system. This shift in food resources is evident on Mainland Southeast Asia.

Saowakontha et al. conducted a study on edible forest products in two villages, Ban Moh and Ban Nong Khong, Phu Wiang district, in Northeast Thailand, presenting a list of 34 wild food plants. They found that the degree of dependency on this resource was related to the distance from the village to the forest, thus, the longer the distance to the forest, the higher the dependency on other areas for food gathering. Likewise, Kosaka et al. compared two rice farming villages from Savannakhet Province, Laos, obtaining the same results.

Whereas Bak village, located in the uplands with an extensive forest area, showed to be more dependent on forest diversity, farmers from Nakou village, situated in the lowlands with a small area of remnant forest, identified more useful plants from the rice fields than the forest, compensating for the lack of resources by maintaining the tree diversity within the paddy rice fields. Studies conducted specifically on non-timber forest products provide surprising results. For example, in a study conducted in the Lao P.D.R., the researchers discovered that farmers used multiple land types and that 60% of the non-timber forest products were not from the forest at all but were collected from fields (paddy, dry grass areas, and fallow), streams and ponds. The same happened when Shibahara conducted research on hunting and gathering in public forests of Roi Et, Northeast Thailand. Although research was focused on forest areas, a major finding was that farmers relied mainly on wild foods from rice fields rather than forests. Shibahara also emphasized that most gathering activities occurred on private land instead of public land. The role of private land in food gathering entitlements among Northeast Thai villagers has been documented by Price.

Given the alarming rate of decrease in forest and wooded areas in Thailand it is becoming increasingly important to also study the wild food plants from anthropogenic areas, as several studies have shown that farmers are becoming more dependent on these places for ensuring their household dietary diversity and food security [8,12,14,45,46].

Somnasang, Rathakette and Rathanapanya found that paddies are a principal place for gathering wild vegetables and fruits in Northeast Thailand. Likewise, Price estimated that farmers gather more from the fields than from any other place. Indeed, rice fields are not only important in terms of rice production but are biologically diverse and multi-resource agro-ecosystems. According to the International Rice Research Institute, paddies possess over 100 useful associated plant species being sources of food, medicine, fibre, construction material, fuel and animal feed.

ANTHROPOGENIC ECOSYSTEMS

Rice fields on the plains of Northeast Thailand and Laos are characterized by having trees in the paddy fields, given their importance for local culture and their socio-economic and ecological functions. Trees are either planted or remnants from a previous forest, which went through different stages of transformation until becoming a rice field during the historical and on-going process of agricultural expansion. The transition point was named "rice production forest" by Takaya and Tomosugi. Vityakon et al. recognize different transitional historical stages of land use change, which they describe at the regional, community, landscape and field level in their article "From forests to farm fields: changes in land use in undulating terrain of Northeast Thailand at different scales during the past century". Prachaiyo also explains this process in his publication entitled "Farmers and forests: a changing phase in Northeast Thailand".

There are a number of studies on the diversity of trees in paddy fields in Northeast Thailand. Grandstaff, Rathakette, and Thomas recorded 54 species of trees and shrubs, 32 of them used as food and/or medicine, growing in the rice fields. Watanabe et al. recorded 16 useful tree species growing in paddy fields in the region. Additionally, Vityakon conducted research on the importance of trees for soil fertility in rice fields. She identified 25 species (14 of them used as food and/or medicine) surviving from previous forests, indicating, if applicable, their uses as food and/or medicine. Later on, Prachaiyo described 28 useful tree species growing in the paddies mainly for timber, latex, food, medicine, oil or fodder. Subsequently, Tipraqsa emphasized the importance of trees in rice fields in Northeast Thailand, documenting 52 trees found in the diverse farming systems in the rice landscape. Finally, trees in rice fields have also been systematically documented in Laos by Kosaka et al., and also discussed in the symposium "Tree-Rice Ecosystem in the Paddy Fields of Laos" organized by a Japanese-Thai project on the same topic, where the utilization of some tree species as food was noted.

Plant diversity in rice fields not only consists of trees, but also aquatic and terrestrial herbs, climbers and shrubs. However, several herbs, climbers and shrubs are classified as weeds or invasive species by agronomists. Yet, a number of weeds are used as vegetables or medicines in Thailand. Maneechote documents 59 edible weeds indicating their parts eaten and the habitat where they grow, which corresponds to about 30% of the 150 plant species classified as weeds in the country. Vongsaroj and Nuntasomsaran conducted a literature review on weed utilization in Thailand reporting 33 weeds used as food, 16 as medicine and 12 as animal feed; some of them were also listed later on in Vongsaroj's. Kosaka et al. identified 11 edible species, 5 medicinal species and 2 plants used as animal feed, mostly weeds from the paddy fields in Savannakhet, Laos.

Prachaiyo also listed some herbs used as vegetable or medicinal plants growing in the rice fields of Northeast Thailand. Although weeds have been shown to have diverse uses around the world, they are continuously overlooked in their role as sources of food and medicine. Minor attention is paid to weed utilization in Thailand given that most agricultural research is focused on minimizing their population.

THIS STUDY

Despite the recognition of the important role that wild food plants play for farmers' livelihoods in Northeast Thailand, information is rather scattered throughout different publications, which are mainly in the Thai language. There is no single article presenting not only an exhaustive list of species but also their local name and, botanical and cultural characteristics. This is certainly necessary as a baseline for future research in this area.

The objectives of this paper are to provide selected results from an ethnobotanical study of wild food plants conducted in Northeast Thailand. A complete botanical inventory of wild food plants used by the study villages and their surrounding areas is provided including their diversity of growth forms, the different anthropogenic locations were these species grow and the multiplicity of uses they have. The research presented in this paper contributes to understanding the importance of different anthropogenic ecosystems where wild food plants grow and provides insights on the multiplicity of uses of these plants.

METHODS

TAXONOMIC IDENTIFICATION AND PLANT NAMING

Fieldwork was conducted from 2006 to 2010, taking as a baseline the results obtained in research carried out by one of the authors in two adjacent villages located in Kalasin Province, where she identified 77 species classified as 'wild' food plants during focus group elicitations conducted with local farmers. This list was built upon and increased using focus groups and key informant interviews as complementary methods in the same villages. A final list of 87 species of locally classified 'wild' food plants was constructed and local names of plants in the local Thai-Lao vernacular were recorded in the Thai script. Species were botanically identified by taxonomists from the Department of Biology of Chang Mai University and Walai Rukhavej Botanical Research Institute of Mahasarakham University. Herbarium specimens of most of the identified species are on repository in one or more locations in Thailand, including the Bangkok Herbarium of the Department of Agriculture (BK) in Bangkok, Herbarium of Walai Rukhavej Botanical Research Institute (WRBG) in Mahasarakham, and the Herbarium of Khon Kaen University (KKU) in Khon Kaen. Botanical naming of family, genus and species follows "Flora of Thailand".

The villagers use the term geht eng, which means "birth itself" for wild food plants. However they do distinguish between "birth itself" as a type of plant versus just the verb "to birth by itself" (without human intervention, such as sowing or transplanting). Some "birth itself" species can also be transplanted or propagated, as some domesticates such as tomatoes can "birth themselves" (growing from consumption debris). Domesticates that "birth themselves" are not considered wild food plants ("birth itself" type of plant). Plant types are further identified by prefixes.

The most common prefixes used for naming food plants refer to their edible part, such as bak and maak that mean fruit yod meaning shoots bai (which is a more unusual prefix) referring to leaf and dok that means flower A very common prefix for naming wild food plants isphak which means vegetable. Phak includes shoots, leaves, stems and sometimes whole aerial parts eaten as vegetable. In this way, if a plant has more than one edible part, it will likely have more than one name differing in the prefix used. For example, Garcinia cowa has two local names Phak moong and Bak moong given that it is eaten as both vegetable and fruit.

A total of 131 plant names were documented for the 87 plants, giving an average of 1.5 names per plant. Plant names were carefully recorded in the local Isaan dialect (capturing both pronunciation and local tone) using the Thai script. Plant names were also transliterated into English. Finally, English names were obtained from Germplasm Resources Information Network (GRIN), Multilingual Multiscript Plant Name Database (MMPND) and Plant Resources of Southeast Asia (PROSEA).

ETHNOBOTANICAL DATA COLLECTION

Growth form and life cycle were determined for each species through field observation and complemented with literature. Growth location and cultural characteristics of the plants, such as edible parts and multiple uses, were assessed through focus groups and supplemented with key informant interviews conducted not only in the research villages but also in two additional nearby villages. The use of different methods permitted, to a certain degree, triangulation and greater depth. These activities were carried out with the aid of local translators who speak the Thai-Lao vernacular of the Lao language (Isaan) as it is spoken in the research location and are knowledgeable about the research topic. Finally, a relational data base of wild food plants was built using MicrosoftAccess.

Focus groups are particularly useful when the everyday use of language and culture of particular groups is of interest, and when one wants to explore the degree of consensus on a given topic. The focus group method has previously been successfully applied to the collection of plant species level information with farmers in Northeast Thailand. Each focus group consisted of six to nine members, following Bernard's recommendations on the number of participants. Focus group participants were generally middle-age women or slightly older (34 to 66 years old), named by the villagers themselves as knowledgeable about contemporarily gathered plants. A total of 12 sessions were carried out sometimes with different participants, each session lasted two to three hours and was tape recorded. All of who participated in the study did so freely and with consent.

RESULTS AND DISCUSSION

BOTANICAL CHARACTERISTICS OF WILD FOOD PLANTS

A total of 87 wild food plants, belonging to 47 families, were mentioned by farmers through key informant interviews and focus group discussions in 2006, building up on a previous list of plants documented by Price in 1990. Out of this total, 76 plants were botanically identified to the species level recognizing a total of 75 different species (two plants correspond to different sub-species of the same species), 9 were identified to genus level and for two botanical identification was not possible. About 13% of the plants were from the Leguminoseae family (6 species belonged to Mimosoideae and 5 to Caesalpinioideae). Other important families were Annonaceae, Myrtaceae, Poaceae, Pontederiaceae, Sapindaceae, Zingiberaceae, with 3 species each.

Two categories of life cycles were considered: annual and perennial. Some 79% of the wild food plants were perennial and 21% annual. For analysing growth form, seven categories were considered: aquatic herb, terrestrial herb, climber, shrub, tree, bamboo and rattan. Other important growth forms were terrestrial herb (18%), aquatic herb (15%) and climber (13%). Shrubs only presented five plants, followed by bamboo with three plants and rattan with only one plant. Climber and terrestrial herbs were both annual and perennial, while aquatic herbs were only annual plants. Trees, shrubs, bamboos and rattans were all perennial plants.

GROWTH LOCATION OF WILD FOOD PLANTS

From an ecological perspective, local farmers provided two major kinds of answers when they were asked where a plant grows. Firstly, (a) they gave general names of what ecologists regard as anthropogenic ecosystems, such as rice field or home garden; and secondly (b) they provided names of specific sub-systems of an anthropogenic ecosystem, such as field margin, tree row or water pond, which all are part of the rice ecosystem. In order to facilitate the analysis, the answers were grouped into six major growth locations: rice field, secondary woody area, home garden, upland field, swamp and roadside, including plants that grow in any of the sub-systems. The analysis of the ethnoecological classification of growth locations (local emic categorization) was not an objective of this paper.

The six major growth locations of wild food plants are the following:

1. Rice field, containing a diverse range of aquatic, semi-terrestrial and terrestrial niches, is where most wild food plants, roughly 70%, can be found. Only six plants out of 61 are exclusively found in the rice fields (mainly terrestrial herbs regarded as weeds), whereas the rest can also be found in other places, mostly home gardens (64%), secondary woody areas (45%), upland fields (40%) and swamps (20%). In rice fields it is possible to find aquatic herbs such as Nymphaea pubescensand Neptunia oleracea; terrestrial herbs such as Limnophila aromatica and Amaranthus viridis; trees as Borassus flabellifer and Leucaena leucocephala; and climbers like Coccinia grandis.
2. Fifty-five percent of the plants occur in secondary woody areas, which are mainly public areas located outside the farms, near upland fields. Only eight out of 48 plants were noted as growing exclusively in woody areas, whereas the rest grow also in other locations, mainly rice fields (68%) and/or home gardens (65%), some of which having been transplanted by the villagers. Most of the wild food plants growing in the woody areas are trees (65%), such as Azadirachta indica (also growing in home gardens and rice fields) and Canarium subulatum (found only in woody areas). A culturally important terrestrial herb only gathered in woody areas is Curcuma singularis, which is gathered in the rainy season.
3. Fifty-two percent of the plants occur in home gardens. There were no plants exclusive to home gardens, all plants could be found in other locations, mainly rice fields (78%), woody areas (58%) and upland fields (49%). Many species growing in home gardens are transplanted from other areas and subject to different degrees of management, such as Tamarindus indica. Species in home gardens are mostly trees (e.g. Phyllanthus acidus) and climbers (e.g. Tiliacora triandra andMomordica charantia), followed by a few terrestrial herbs (e.g. Centella asiatica).
4. Upland fields, mainly consisting of fields with cash crops of cassava and sugar cane, contain 37% of the wild food plant species. No plants were exclusive to the upland fields. Wild food plant species that occur in upland fields also grow in other locations, mainly woody areas (84%), rice fields (69%) and home gardens (69%). Most species are trees such as Syzygium gratum andCareya arborea.
5. Swamps contained 17% of the plants. Three out of 15 plants were exclusive to swamps, but these are rarely found. The rest of the plant species also occur in rice fields, with the exception ofNeptunia javanica which is a terrestrial herb found in home gardens and roadsides.

Regarding their growth form, 75% are aquatic herbs such as Hydrolea zeylanica, while 25% are terrestrial herbs such as Oenanthe javanica.

6. Thirteen percent of the plants grow on roadsides. No plants were exclusive to roadsides. All plants found at roadsides also grow in home gardens. Nine roadside plant species also occur in the rice fields, seven in the upland plantations and six in the woody areas. Most of the wild food roadside plant species were trees such as Pithecellobium dulce and Cassia siamea. There are a few climbers such as Passiflora foetida.

Wild food plants are widely distributed in the anthropogenic landscape. The results show that 80% of the wild food plants can be found in multiple growth locations, particularly rice fields, woods and home gardens. Forty percent of the wild food plants we documented grow in two different locations, 24% grow in three locations and 16% grow in four or more different locations. This can be explained, in part, by species being moved from one place to another facilitated by different degrees of management. This is consistent with the findings of Price and Chanaboon et. al., who reported the presence of wild food plant management practices in Northeast Thailand.

Out of the 38 tree species, 31 (82%) are to be found in the secondary woody areas, 26 (68%) in the rice fields, 22 (58%) in home gardens and 22 (58%) in upland fields. As discussed in the introductory section, the presence of trees is a common characteristic of rice ecosystems in Northeast Thailand. Trees grow in hillocks, shelters, tree rows and pond margins diversifying the habitats and facilitating the presence of climbers and other plants in the fields. Most trees are maintained in paddies due to their use value. For instance, two thirds of the trees are medicinal (66%) and, in addition, some provided timber and fuel.

MULTIPLICITY OF USES, INCLUDING PARTS USED

The edible parts of wild food plants vary from reproductive structures (flowers, fruits, seeds) to vegetative organs (leaves, shoots, stalks of flower, stems and sometimes the whole aerial part is consumed). For somewhat less than half of the plants only one part is edible (47%), e.g. only the shoots of Neptunia oleracea are consumed. More specifically, for 25% of the plants two parts are eaten, which is the case of Adenia viridiflora (shoot and fruit). For 12%, three parts are eaten, such as Senna sophera (shoot, flower and fruit). And for 16% of the plants, more than three parts are eaten as for Limnocharis flava (shoot, flower, stalk of flower and fruit).

In order to facilitate the analysis, eight categories of different parts consumed were established:

- Young shoots sprouting from roots, stems or tips of plants are consumed in 53% of the wild food plants, such as Bambusa bambos,

Senna sophera and Telosma minor. Shoots are widely consumed regardless of the growth form and life cycle of the plant.

- Fruits, which can be eaten unripe and/or ripe depending on the plant, are consumed in 39% of plants, mainly trees and climbers. The fruit of Tamarindus indica is very popular both unripe (it is sour, seasoned with fish sauce and chili) and ripe (it is very sweet, eaten raw or its juice added to a dish of food).
- Flowers or inflorescences are consumed for 24% of plants. Typical species are Dolichandrone serrulata and Curcuma singularis.
- Whole aerial parts, including shoots, young leaves and tender stems, are consumed for 14% of plants. This is the case of many terrestrial and aquatic herbs including Limnophila aromatica andGlinus oppositifolius, with the exception of Cuscuta chinensis that is a climber.
- Leaves, mainly eaten when young and tender as a raw vegetable or cooked in traditional dishes, are consumed for 9% of plant species like the climber Cassytha filiformis and the treeLeucaena leucocephala.
- Seeds are consumed for 7% of plants. For example, the seeds of Irvingia malayana are eaten roasted as a snack.
- Stalks of flower or inflorescence are eaten in the case of 6% of the plants, including Nymphaea pubescens whose stalk is eaten raw as a side dish.
- Stems are consumed for 5% of the plants, including the edible stems of the aquatic herbLudwigia adscendens, the inner core of the trunk of the tree Borassus flabellifer (used to make sweets), and the rhizomes of the terrestrial herb Alpinia malaccensis.

More than two thirds of the wild food plants presented other uses besides food (71%). Some 35% of plants had one additional use, while 26% of the plants had two additional uses, 7% had three additional uses, and three plants had four or more additional uses.

- Medicine was the most widely mentioned additional use (60% of the plants). Moreover, it is remarkable that out of the 30 plants with an additional use, 28 have medicinal uses. Some examples of medicinal plants are the herbs Centella asiatica and Ludwigia adscendens.
- Fodder use was reported for 16% of the plants. More than half of these fodder plants (9 plants) are also regarded as medicine, such as Leucaena leucocephala and Coccinia grandis. Fodder plants are mostly herbs, trees and bamboos.
- Twelve percent of the plants are used as fuel, like Nephelium hypoleucum and Cratoxylum formosum. Plants used as fuel were

mainly trees growing in the rice fields and home gardens, many of them are also found in the woody areas.

- Timber was reported for 8% of plants. It included trees such as Xylia xylocarpa and Spondias pinnata.
- Eight percent of the plants are used for making local handicrafts. The three bamboo plant species are typically used in handicraft production such as in weaving hang mats. The wood ofArtocarpus lacucha is used to make a traditional musical instrument similar to a xylophone calledpong lang, which is regarded as the symbol of Kalasin Province.
- Domestic use was reported for 6% of plants. For example the rattan Calamus sp. is used for making home utensils.
- Five percent of the plants have auxiliary uses. The leaves of Azadirachta indica are utilized to make natural insecticide. Leucaena leucocephala (Leguminoseae) is used as fertilizer. All four plant species are also used as medicine.
- Ritual use was reported for 3% of the plants. The Buddhist monks spread holy water using the leaves of Phyllanthus acidus. Villagers make curry with the young leaves of Aegle marmelos and give it to the monks in blessing ceremonies.
- Dye was mentioned for 3% of plants used as natural colorants. The fruit of Tamarindus indica is used as dye for fish nets. The bark of Cratoxylum formosum is utilized to dye clothing.
- Two plants are used for cleaning, for example Cassia siamea is used for making shampoo.
- Only one plant is used for chewing. The bark of Artocarpus lacucha is chewed, sometimes with betle nut.

Consistent with the findings of Price for Northeast Thailand, the importance of wild food plants as food-medicines is present in the current findings. The results indicate that these wild food-medicine plants are important not only for their curative properties, but also for their nutritional and preventive properties. Indeed, this overlapping role as a source of both food and medicine has been documented for farmers' use of wild plants in numerous parts of the world.

For example in Vietnam, among the Hausa of Northern Nigeria, among Albanians and Southern Italians in Lucania, in the North West Bank, Palestine, and in the Inner Mongolian Autonomous Region, China. Furthermore, undoubtedly, there is an overlap of food, medicine and animal feed, given that almost two thirds of fodder plants are also medicinal (9 out of 14 fodder plants). These results seem to follow the pattern of Ogle et. al. who discussed the multiple functions of wild food plants in Vietnam.

Villagers also mentioned additional uses of wild food plants related to the ecological services they provide. For instance, they commented that the aquatic herb Monochoria hastata, which is regarded as a weed of rice fields, provides shade for fish. Additionally, many trees were acknowledged as habitats of red ants and other edible insects. Fish and insects, among other animals, are also gathered from the rice fields constituting an important part of the local diet.

CONCLUSIONS

This study shows the remarkable importance of anthropogenic ecosystems in providing wild food plants. This is reflected in the great diversity of plants found, contributing to the food and nutritional security of rice farmers in Kalasin, Northeast Thailand. The data compiled in this study shows that the majority of wild food plants grow in the different aquatic, semi-aquatic and terrestrial sub-systems offered by rice agro-ecosystems. Trees presented more plants than other growth forms, constituting an important feature of different terrestrial sub-systems of the paddies, such as hillocks, tree rows and shelters. Many important plants are aquatic and terrestrial herbs, as well as climbers. Both annual and perennial species are present in significant numbers.

One of the main findings is that most wild food plants are found in multiple locations, and more than half of them grow either in rice fields and home gardens, rice fields and woods, home gardens and woods, or rice fields, home gardens and woods. No plants were exclusive to home gardens and very few plants were exclusive to woods and rice fields. From these results we assert that farmers play an active role in managing many of these plants, for example, transplanting them from the woods to the fields or to home gardens, making them available in those anthropogenic places located closer to their house and village. This assertion follows the patterns proposed by the "botanical dietary paradox", which clarifies the use of so many wild food plants by farmers in that when deforestation occurs, farmers change to gathering new wild food plants closer to home, including the weeds of agriculture.

Another major point to note from the results of this research is that more than half of the wild food plants have many edible parts, and more than two thirds of them have additional uses. Shoots, sprouting from the tips of plants, stems or roots, were the most widely cited as consumed regardless of the growth form or life cycle of the plant. Fruits were also common, particularly collected from trees and climbers. Wild food plants presented more than eleven additional uses, accentuating their overall relevance for rice farmers. The most common additional use was for medicine.

The data compiled in this study highlights the necessity to better understand the role of anthropogenic ecosystems in providing wild food plant resources. Further research needs to be carried out on the seasonal quantification of their environmental availability, as well as the location of actual gathering events. Finally, research on transplanting and other management practices would allow us to better comprehend the distribution of these plants in the different ecosystems.

ETHNOBOTANY AND ECONOMIC BOTANY OF THE NORTH AMERICAN FLORA

When Europeans first arrived, in both eastern and southwestern North America north of Mexico, they found people who were practicing agriculture, much of it with crops from Mexico. Consequently, the use of native wild plants received scant attention. This changed, however, when the Europeans penetrated the areas inhabited by hunters and gatherers. According to R. I. Ford (1986), "the traditional use of plants and animals by American Indians is better documented than for the early peoples of any other continental area of the world." Ford has brought together a number of the significant papers dealing with the use of plants and animals by the native people. Furthermore, archaeological investigations, particularly in the last half century, have also contributed greatly to our understanding of the plants used by the native North Americans.

The immigrants to North America from Europe brought the Old World crops to North America, and those plants soon came to be the dominant cultivated crops in northern North America. Many weeds, a few of which were found to serve useful purposes, were also introduced unintentionally.

WILD FOOD PLANTS

The American Indians utilized a large number of plants for food or beverages. E. Yanovsky (1936) listed 1112 species as so used. Many of these plants were of minor use, and a few were not natives but introduced species. The vast majority of the plants in the list are angiosperms, of course, but algae, fungi, lichens, seedless vascular plants, and gymnosperms are also represented.

One of the most widely distributed sources of food was provided by oaks (Quercus spp.). Acorns of over 25 species were used, with those of various western species providing the basic food for some California Indians. Various ways were devised to remove the bitter principles to make the acorns palatable. Nuts of other genera, such as chestnut (Castanea spp.), hickory (Carya spp.), and walnuts (Juglans spp.), were also widely used.

Camas (Camasia quamash) was widely used in western North America. The bulbs were cooked in various ways and then eaten or ground and made into a flour that was used for making a "bread." In the central and eastern area, the tubers of the potato bean (Apios americana) were cooked for food.

Fruits or "berries" belonging to various families were gathered throughout most of North America. Rosaceae provided the greatest number of contributions. Grains of a number of grasses were collected for food. One of the grasses, wild rice (Zizania palustris), was a staple in the Great Lakes region. The tapping of maple (Acer saccharum and other species) for sugar was practiced in the central and eastern areas as it still is today. Less widely known is that the pounded bark of these species was also used for food. Among the gymnosperms, the edible seeds of the pinyon (Pinus edulis) and of several other western pines stand out as important food sources.

Many of these wild plants were used by the Old World immigrants in the early historical period, and even now some people collect wild foods, but more as a hobby than as a necessity. Many books, often regional in their treatment, such as M. L. Fernald et al. (1958), list edible plants and give recipes for their use. A number of the plants gathered from the wild at present are introduced weeds or escapes from cultivation, including favorites such as asparagus (Asparagus officinale), burdock (Arctium lappa), chicory (Cichorium intybus), wintercress (Barbarea vulgaris), daylily (Hemerocallis fulva), and watercress (Rorippa nasturtium-aquaticum).

DOMESTICATED FOOD PLANTS

The archaeological record indicates that food production was being practiced in northern North America over 3000 years ago. The principal food crops in eastern North America at the time of contact with Europeans were corn, also called maize (Zea mays), squash (Cucurbita pepo), and beans (Phaseolus vulgaris). All of these plants are known much earlier in archaeological records from Mexico, and for some time it was naturally assumed that food production came to northern North America, along with these plants, from Mexico. The possibility that there had been an independent domestication of plants in northern North America was mentioned as early as 1924 (B. D. Smith 1987), and for some time evidence has indicated that the sunflower (Helianthus annuus) and sumpweed (Iva annua) were domesticated in eastern North America before the arrival of corn and beans from Mexico.

More recently B. D. Smith (1985) has shown that a domesticated chenopod (Chenopodium berlandieri or C. bushianum) was also present in eastern North America. Other plants, such as May grass (Phalaris caroliniana), were being cultivated, but there is no evidence that they were domesticated (C. W. Cowan 1985). When the Europeans arrived, the sunflower was seen in cultivation, but both Iva and Chenopodium had apparently already disappeared as domesticated plants. Two rather cryptic references in the literature, however, might indicate that Chenopodium was still cultivated at the time Europeans settled the continent (D. L. Asch and N. E. Asch 1977).

The hypothesis that plants were domesticated in eastern North America prior to the introduction of Mexican crops seemingly became untenable when both squash (Cucurbita pepo) and bottle gourd (Lagenaria siceraria) were reported at a site in Missouri dated at over 4000 years old (M. Kay et al. 1980), because it was thought that both of these plants came from Mexico where they were cultivated at much earlier dates. It now seems likely, however, that C. pepo was a natural element in northern North America and that it is still represented there by the plant usually known as C. texana. Therefore it is possible that C. pepo was domesticated independently in Mexico and farther north in North America (D. S. Decker 1988; C. B. Heiser 1989). The bottle gourd cannot be considered an indigenous plant of North America; its entry could have been as a weed, with or without human aid. The possibility even exists that it came to Florida by ocean currents from South America (C. B. Heiser 1990).

In addition to the plants mentioned previously, the Jerusalem artichoke (Helianthus tuberosus) was also cultivated in eastern North America. Because no archaeological material is yet known, it is impossible to assign any dates to its domestication.

In the Southwest about 3000 years ago, agriculture began with what has been designated as the Upper Sonoran agricultural complex (R. I. Ford 1985b). Maize (corn), squash (Cucurbita pepo), beans (Phaseolus vulgaris), and the bottle gourd, all of which apparently came from Mexico, were represented. Some 1500 years later the Lower Sonoran agricultural complex developed. In addition to the plants from the earlier agricultural complex, it included other squashes (Cucurbita mixta and C. moschata), tepary bean, lima bean (P. lunatus), and jack bean (Canavalia ensiformis), and a panic grass (Panicum sonorum). These plants all could have come from Mexico, but possibly some of them, such as the tepary bean and the panic grass, were also domesticated in an area that is now within the boundaries of the United States.

Following the settlement by Europeans, the agriculture of North America soon underwent a drastic change as crops brought from the Old World spread. The present dependence of North American agriculture on crops of foreign origin is indicated. Of the food crops, only the sunflower is indigenous to North America north of Mexico, and only two of the others, maize and beans, were being grown in northern North America at the time of the discovery. Seven other crops of those listed ---potato, sweet potato, tomato, peanut, avocado, tobacco, and cotton---are also native to Latin America; all of the others are Old World in origin.

Today the United States is the leading country of the world in the production of maize (corn), oats, sorghum, tomatoes, soybeans, and peaches, and second for six other crops. Canada is first in the production of linseed and second for rapeseed.

NARCOTICS, HALLUCINOGENS, STIMULANTS, AND ALCOHOLIC BEVERAGES

Tobacco, usually smoked, was widely used in northern North America. The domesticated Nicotiana rustica was cultivated in the East in prehistoric times and in the Southwest in the early historical period. In other regions tobacco from several native species was gathered from the wild, or occasionally was cultivated by the Native Americans. All of these tobaccos were replaced by Nicotiana tabacum, which was introduced from Latin America by John Rolfe in 1610 or 1611 and soon became the basis of the tobacco industry in Virginia and, later, in the other colonies (C. B. Heiser 1969).

Although many plants were used as hallucinogens in Mexico, only a very few were employed in northern North America. Species of Datura were employed prehistorically in ceremonies in the Southwest. Datura inoxia, used in Arizona, New Mexico, and California, was apparently the major species. Seeds of the mescal bean, Sophora secundiflora, a species found naturally in New Mexico and Texas, were also used prehistorically as a hallucinogen; a cult involving their use developed in the Plains area in modern times. Peyote (Lophophora williamsii), which has its northernmost limit in the lower Rio Grande valley of Texas, was used by Indians in Texas as a hallucinogen in 1760. Its use among Native Americans began to spread about 1880 and reached Canada in this century. Peyote has been legalized for use in the Native American Church. More details on these plants may be found in R. E. Schultes and A. Hofmann (1980). Today, the introduced marijuana (Cannabis sativa) is the most widely grown hallucinogen in North America. Because it is illegal, exact production figures are not available.

Leaves of yaupon, Ilex vomitoria, were the principal ingredient of a drink used by the Indians of the Southeast to induce vomiting as part of a purification ritual (H. E. Driver 1961). A tea was also made from this and other species of Ilex (E. Yanoksvy 1936). Ilex is known to contain caffeine.

The Indians of the Americas never distilled alcohol from plants; however, naturally fermented beverages were utilized in some places. Fruits of various cacti were sometimes fermented by Indians in Arizona to make a wine (H. E. Driver 1961).

Unlike the Indians in many other parts of the Americas, those of northern North America did not adopt maize (corn) as a source for an alcoholic beverage before the conquest. This was remedied sometime later by people of European descent who used maize to prepare a whiskey that became known as bourbon (after Bourbon County, Kentucky), and this is the only distinctive important contribution of northern North America to the world's alcoholic beverages. A beer from persimmon (Diospyros virginiana) was reportedly made by Indians in the East (U. P. Hedrick 1950). The early colonists tried to make wine from the native grapes in eastern North America without much success. The Old

World grape (Vitis vinifera) was successfully introduced into California by the Spanish and became the basis for the development of the wine industry in the United States.

MEDICINES

The native North Americans used more plants for medicines than they did for food. D. E. Moerman (1982) lists 2147 species as employed medicinally. Some of the same plants were used, of course, for both food and medicine. Many medicinal plants of Native Americans were later adopted by the European colonists (V. J. Vogel 1970), who also made medicinal use of introduced species that they had brought with them. C. F. Millspaugh (1892) described 180 plants as so used. T. Arnason et al. (1981) have provided a recent treatment of the plants used for medicine in eastern Canada. C. Prescott-Allen and R. Prescott-Allen (1986) reported 67 active ingredients from wild plants as occurring in 10 or more drug products sold in Canada; about one-fourth of the plants are native only to North America.

The principal wild plant of northern North America in terms of number of drug products produced from it is cascara sagrada from the bark of Rhamnus purshianus, native to the northwestern United States and British Columbia. In 1977 the annual retail value of cascara sagrada in the United States was 75 million dollars; it is generally thought to be the world's most widely used cathartic.

A second medicinal plant that deserves particular attention is ginseng, Panax quinquefolius, of eastern North America. Although of limited importance to Native Americans, in 1980 some 575,000 pounds of roots with a value of 39 million dollars were exported. About 85% went to Hong Kong. Approximately 70% of the exported roots were derived from cultivated sources. Destruction of its habitats and overcollecting have contributed to a decline in the supply of the wild plants, whose roots are considered more desirable than those of cultivated plants.

WOOD

Forests, much of them coniferous, once covered much of North America. Trees (as sources of timber and, especially, as sources of fuel) were, in fact, northern North America's most important raw material until the twentieth century. The United States is still the world's leader in timber production by volume, and Canada is sixth; by value Canada is the world's top timber producer, and the United States is second (C. Prescott-Allen and R. Prescott-Allen 1986).

Wood is consumed in four main ways: as fuel, as pulp for the manufacture of paper, for construction (including furniture), and for by-products, such as rosin, turpentine, and terpenes. The use of wood for fuel has declined greatly over the years and now only about 10% of it is burned, whereas its use for pulp has greatly increased, over 30% now being so used. Nearly 60% is used for construction.

Silviculture has slowly taken hold in North America. About 800,000 hectares have been planted to trees each year in the past three decades in the United States. Most of the planting is of soft woods, principally pines (Pinus spp.). The principal hardwoods planted on a commercial scale are cottonwood (Populus spp.) and black walnut (Juglans nigra). Much of the farmland in the southeastern United States is now devoted to production of timber and/or wood for pulp.

OTHER EARLY USES OF PLANTS

Native Americans used many wild plants both for dyes and for fibers. Species of Apocynum, Urtica, and Linum were most widely used for their fibers. Cotton (Gossypium hirsutum) was the only domesticated fiber plant in prehistoric times. It came to northern North America from Mexico. In precontact times, its cultivation within the flora was limited to a small area in the Southwest. In historical times, of course, cotton became widely cultivated in the southern United States. The devil's-claw (Proboscidea parviflora), used for a fiber in basketry, was domesticated in the Southwest. Its domestication, however, is apparently rather recent (P. K. Bretting and G. P. Nabhan 1986).

ORNAMENTALS

Plants from northern North America are now used for their ornamental value in landscape plantings or in gardens both in North America and other parts of the world. Many woody species, often little or not at all changed from the wild types, are planted, and some, such as Magnolia grandiflora, have particularly extensive distributions as cultivated plants. A number of native and horticulturally unchanged herbaceous angiosperms and ferns are used in wildflower gardens, and cultivars of many of the native species are offered in the trade for use in more formal gardens. Some species, such as those of goldenrod (Solidago spp.) and sunflower (Helianthus spp.), are more appreciated in Europe than in their homeland. Descriptions of many of the native North American plants grown as ornamentals may be found in Hortus Third.

NEW AND INCIPIENT DOMESTICATES

Certain plants consumed by Native Americans for food have become domesticated only in the past century or so; these include the pecan (Carya illinoinensis), the blueberry (Vaccinium corymbosum), the cranberry (V. macrocarpon), and most recently, wild rice (Zizania palustris). According to U. P. Hedrick (1950), 11 kinds of plums (Prunus spp.), 15 kinds of grapes

(Vitis spp.), 6 species of blackberries (Rubus spp.), and 4 species of raspberries (Rubus spp.) have also become commercially valuable cultivars derived from native species. Research is proceeding with several other native plants, particularly those for cultivation in semiarid regions, and particularly for products that might have industrial uses (e.g., jojoba [Simmondsia chinensis] for oil, buffalo gourd [Cucurbita foetidissima] for starch and oil, guayule [Parthenium argentatum] for rubber, and gumweed [Grindelia camporum] for resin). Some of these have already proved to be of value.

WEED BIODIVERSITY AND RICE PRODUCTION DURING THE IRRIGATION REHABILITATION PROCESS

Paddy fields provided various ecosystem services, such as (1) provisioning of rice grain, rice straw, and other plant species (i.e., weeds) and animals; (2) regulating air temperature and flood control; (3) cultural services of festivals and rituals associated with farming; (4) supporting nutrient cycles and disease control; and (5) preserving genetic diversity. Paddy fields were traditionally characterized by a rich flora, with more than 1800 plant species (both aquatic and terrestrial) associated with rice in southeast Asia, cropped using traditional methods, i.e., water flooding but with distinct ecological phases, incomplete leveling, heterogeneous microscale conditions, repetitive disturbance due to tillage, low fertilizer input, hand weeding without using synthetic herbicides.

Cambodian paddy fields might be unique in that relatively few farmers adopted modern agricultural techniques, partly due to the country's long civil war (from 1970 to 1993). For example, the rate of fertilizer use was much lower than that in neighboring countries: 5 in Cambodia versus 133 in Thailand and 324 kg/ha in Vietnam, based on averages from 2002 to 2004 (Yu and Fan, 2009).

The area devoted to dry-season fully irrigated rice was only 14% of the total cultivated area in Cambodia (JICA, 2010). The Cambodian government recently issued a national policy to increase rice production from 7.3 in 2010 to 9.1 Mt in 2015 by increasing the dry-season irrigated rice area from 385,000 in 2010 to 480,000 ha in 2015; the government also aimed to increase exports of unpolished rice from the estimated value (including unofficial transportation to its neighboring countries) of 2.06 in 2010 to 2.89 Mt in 2015 (RGC, 2010). In the future, Cambodia's rich weed biodiversity might decrease in those areas where modern agricultural techniques were widely introduced, such as paddy fields undergoing irrigation rehabilitation.

The aim of this study was to provide baseline data of weed flora for the assessment of possible trade-offs and synergies between weed biodiversity and rice production in Cambodian paddy fields under a scenario of shifting management schemes from traditional to modern methods. A hypothesis was proposed that the rice ecosystem was heterogeneous and level of production intensity was relatively low, which might allow up to some extent to increase rice yield without decreasing weed diversity and weed provision function of ecosystem services in the Cambodian paddy.

MATERIAL AND METHODS

Cambodia had a tropical monsoon climate, with a dry season that lasts from December to May and a rainy season that lasts from June to November. The monthly maximum and minimum temperatures in Battambang province were 33.0 (in April) and 23.2 °C (in January), respectively, and the annual precipitation averaged 1368 mm.

The Kamping Puoy Irrigation Rehabilitation Area (KPIRA), located west of Battambang city in Battambang province of northwestern Cambodia (13°02' N, 103°04' E), was one of the largest irrigated paddy areas in Cambodia (ca. 2800 ha out of the ca. 5000-ha area were irrigated in 2008). The irrigation water came from the Kamping Puoy reservoir through irrigation canals, which were constructed during Pol Pot's regime from 1975 to 1979. The irrigation system did not function well, possibly due to poor design, and it was repaired by Italian and Japanese government agencies from 1998 to 2006 (Try, 2008). The goal of the rehabilitation project was to maximize the area that could be planted with rice during the dry season and hence to improve the system to permit cultivation of two rice crops.

The study site was the downstream 950-ha zone of KPIRA, located about 12 km from the Kamping Puoy reservoir. The irrigation system in the 950-ha zone was rehabilitated by the grassroots Kusanone scheme of the Japan International Cooperation Agency (JICA) from 2001 to 2003, and dry-season rice production was only partly implemented between 2003 and 2008.

In the 950-ha zone, six secondary irrigation canals and six secondary drainage canals (ca. 1–3.5 km long) were connected to the main irrigation and drainage canals. Twenty-eight paddy fields were selected from 140 fields along two secondary canals (D2-1 and D2-7), with D2-1 located about 3.4 km upstream of D2-7 along the main canal (N2). The main and secondary irrigation and drainage canals were not lined with concrete. The paddy fields were grouped into upstream (U), midstream (M), and downstream (D) paddies based on their distance from the main canal along the secondary canals D2-1 and D2-7: D2-1 upstream (1U; 0.3 km, 4 selected from 31 fields), D2-1 midstream (1M; 1.2 km, 6 selected from 38 fields), D2-1 downstream (1D; 2.4 km, 6 selected from 29 fields), D2-7 upstream (7U; 0.1 km, 4 selected from 8 fields), D2-7 midstream (7M; 0.8 km, 4 selected from 20 fields), and D2-7 downstream (7D; 1.6 km, 4 selected from 14 fields).

The paddy areas along D2-1 were rehabilitated earlier than those along D2-7. By 2008, farmers only grew a single crop per year in 7D, 7M, 1D, whereas two crops were grown annually in the other fields. Environmental variables, rice and weed growth variables and weed species were collected in two quadrats per field (each 2 m × 2 m) in 28 fields at 2-month intervals from August 2008 (planting time during the rainy season) to February 2009 (after harvest), for a total of 224 quadrats, while agronomic management variables were collected for each field.

The quadrats were established toward the middle of each field to minimize the influence of edge effects, and their locations were determined by means of GPS (eTrex Legend HCx; Garmin Corporation, Olathe, KS, USA). The same quadrats were used for the five observations from August to February. Mean and standard deviation for all the environmental, rice and weed growth, and management variables were calculated for each of the six paddy groups (1U, 1M, 1D, 7U, 7M, 7D) from the measurements of quadrats or fields.

Environmental variables

As an indicator for water availability and microtopography, the depth of the standing water was determined in all quadrats. Soil moisture was measured by a CS620 HydroSense water content probe and soil hardness by a soil compaction meter (Fujiwara, Tokyo, Japan). Water and soil pH were measured by a Twin B-212 pH meter (Horiba, Kyoto, Japan). In each quadrat, the maximum height of rice plants and weeds was measured using a pole. The coverage of rice and weed within each quadrat was visually assessed by Braun-Blanquet (1964)scale: + = <1, 1 = 1–5, 2 = 6–25, 3 = 26–50, 4 = 51–75, and 5 = 76–100%.

Agronomic management variables

The owners and managers of the 28 surveyed paddy fields in Ta Kream and Poy Svay villages were interviewed about their management practices, including planting methods (transplanting or direct seeding), organic and inorganic fertilizer types and application rates, herbicide use (yes or no), hand weeding (yes or no), traditional weed control method (i.e., mid-season tillage) (yes or no), crop residue management after harvest (i.e., burning) (yes or no), and rice yield. Rice yield was also estimated from the samples from each field to confirm agreement with the interviewed values.

Weed data analysis

Voucher specimens were collected for each plant species, as well as for unidentified species with low frequency and low coverage. Identification of species was based on Dy Phon (2000), Harada et al. (1996), and Kadono (1994) using specimens at the Bangkok Forest Herbarium.

All vegetation data were converted into mean cover classes to improve the normality and homogeneity within groups (McCune and Grace, 2002). The dominance values in the Braun-Blanquet scale were therefore converted into the following mean cover classes: + = 0.5, 1 = 3, 2 = 15, 3 = 37.5, 4 = 62.5, and 5 = 87.5%.

The life forms of the species were divided into five categories based on Reimer (1984): (1) emergent graminoids (e-g), whose roots and basal portions grow beneath the surface of shallow water but whose leaves and stems were primarily in the air and that were grasslike herbaceous plants with leaves that were mostly narrower than their length (i.e., linear in outline); (2) emergent broadleaf plants (e-b), which were emergent plants with leaves about as broad as they were long, including ferns; (3) floating-leaf plants (fl), whose leaves float on the water surface but whose roots were anchored in the substrate; (4) submerged plants (s), which spend their entire life cycle (except during flowering) beneath the surface of the water; and (5) levee plants (l), which prefer upland conditions and therefore appear mainly on levees and were not common inside paddies.

To assess weed diversity of each plant community, species richness (the number of species present in a quadrat) and Simpson (1949) diversity index (D') were used. Correlation analysis was conducted for species richness and Simpson's diversity index with all the other managemental and environmental parameters using all the quadrat data in each month.

Detrended correspondence analysis (DCA; Hill and Gauch, 1980) was performed to investigate the relationships between weed communities and abiotic or anthropogenic factors, some of which were discrete variables with only two categories (e.g., yes or no for the use of herbicides). After omitting rare species that had been recorded four or fewer times, a data matrix was prepared consisting of the 224 quadrats cross-referenced against the 52 species out of the total of 76 species that were identified in these plots. The calculations were performed using all the data for four months together by version 5.10 of PC-ORD for Windows. As was often the case in DCA analysis, Spearman's rank correlations were calculated between DCA scores (i.e., axis 1 and axis 2 which had the highest eigenvalues) and key environmental, vegetation, and rice management variables to characterize the axis.

UTILIZATION OF PLANT SPECIES BY FARMERS

Group discussions with several key villagers (e.g., the village chief) were organized at the three villages (Ta Kream, Poy Svay, and Ta Ngen) in the Ta Kream commune during the dry season of 2009 to learn which weeds farmers collected for use from their paddy fields. The names of weeds (local names, English names, and scientific names, if possible), peak collection season, weed abundance, and popularity among farmers were clarified, although a few weeds were left unidentified. "Popular" means some farmers know the usage, while "very popular" means most of the farmers use the weeds. In the same three villages, 44 farmers completed a questionnaire during the rainy season of 2009 to clarify which plant species were commonly collected in the paddy ecosystem, including the purpose for collecting each. Sources of forage for feeding cattle were also identified to assess the relative importance of provision service deriving from weeds.

RESULTS

Standing water was deeper in downstream fields than upstream during the rice growing period in the rainy season, and it was deeper in canal D2-7 than in D2-1. The depth of the standing water in the paddy fields in October was greatest in field 7D (i.e., 43 ± 21 cm; mean ± standard deviation) and shallowest in 1U (i.e., 7 ± 5 cm). Field 1U also had the earliest recession of standing water, which occurred in November. Due to the slower drainage from D2-7, the soil moisture content in D2-7 fields was higher than that of D2-1 in February (early dry season).

In general, the area of a single field (i.e., one paddy field bounded by a levee) was larger in midstream and downstream fields than in upstream (except for 7D). Rice yield was higher in 1U and 1M, and it was lower in 7U due to a flood that damaged the rice plants. Rice height was greater in midstream and downstream fields, particularly 7M and 7D, compared with upstream. In general, the weed height was lower than the rice height, but some exceptions were noted, such as tall emergent broadleaf weeds in 1D (e.g., Sesbania bispinosa, Alternanthera sessilis, Scirpus pilosus, andScirpus grossus). Weed cover in October was largest in 1M and 1U and smallest in 7M and 7D. In 1U, the total weed cover remained consistently high until February, whereas it decreased sharply after October or December in the other field groups. In 1M, the weed heights in February were lower than those of other field groups.

Organic N from manure application was higher in 1U while it was very low in 1D, 7U and 7M. The N and P inorganic fertilizer rates were highest in 1U and 1M. The majority of farmers used the herbicide 2,4-D, but at a low application rate (0.4 ± 0.2 kg a.i./ha). Mid-season tillage was more widely practiced in 1D, 7M, and 7D than in the other field groups. In 1M, almost 80% of the area was burned by February, after the rice harvest. Rice was established mainly by direct seeding along both canals, but rice with an earlier maturity (by more than 1 month) was also cultivated by transplanting in the upstream paddies (e.g., 1U). The soil of 1U fields was hardest in February.

WEED SPECIES

A total of 76 species in 21 families were identified in the KPIRA paddy fields from August 2008 to February 2009. The dominant families were

Cyperaceae (19 spp.), Poaceae (14 spp.), and Scrophulariaceae (8 spp.). The dominant life forms were emergent graminoids (30 spp.), emergent broadleaf plants (21 spp.), levee plants (13 spp.), submerged plants (9 spp.), and floating-leaf plants (3 spp.).Fimbristylis miliacea was dominant in all the field groups, particularly during the rice growing period. The mean cover of F. miliacea in August was higher in 7M, 1D, and 7D, where dry direct seeding was practiced.

Weed species composition differed between 1U and 7D. The vegetation of 1U was characterized by many members of the Poaceae (e.g., Pseudoraphis spinescens,Echinochloa colonum f. viviparum, Isachne confusa) and Cyperaceae (e.g., S. grossus,Cyperus rotundus), which often appeared on the levees. Species composition in 1U did not change clearly from August to February, and emergent graminoids appeared with high frequency in all months. Marsilea crenata dominated the plots in 1U. In contrast, the indicator species of 7D were common in naturally flooded lands and ponds. In addition, wild rice (Oryza rufipogon) occurred at high frequency in December in 7D. The number of emergent broadleaf plants was largest in 7D. In 7M Limnophila heterophylla was found as indicator species.

Table.Spearman coefficient of Simpson's diversity index (D') and species richness with managemental and environmental parameters in each month.

	August	October	December	February
D′	Species richness (0.47), herbicide use (0.43), midseason tillage (0.36),*weed coverage(−0.29)*	Species richness (0.54), rice coverage (−0.36), growth duration (0.31), organic N (−0.40),*inorganic N (−0.31),* midseason tillage (0.39), distance (0.32)	Rice height (0.34), weed coverage (−0.45), growth duration (0.36), midseason tillage (0.35),*soil hardness (−0.27),water depth (0.30)*	*Species richness(0.27)*, rice height (0.39), *soil moisture content (0.33), weed coverage (−0.32),growth duration (0.29),midseason tillage(0.29)*
Species richness	D′ (0.47), weed height (0.34), rice coverage (0.50), weed coverage (0.37), *rice yield (0.30)*	D′ (0.54), inorganic N (−0.43), herbicide (0.36), midseason tillage (0.40)	Rice height (0.35), weed height (0.40), rice coverage (0.38), distance (0.50), midseason tillage (0.38),*weed coverage (0.27),rice yield (0.34), growth duration (0.32)*	*D′ (0.27)*, soil moisture content (0.41), rice height (0.36), weed height (0.56), rice coverage (0.61), weed coverage (0.60), distance (−0.35), burning (−0.44)

Species richness and diversity were generally higher during the rice growing period (October, December) and higher in downstream fields (e.g., 1D, 7D). In 1D the highest species richness (9.8 species per quadrat) and the highest diversity (D' = 0.46) occurred in October. The post-harvest species richness in February was low in 1M after the burning of crop residues, as well as in 1D due to coverage by a lot of rice straw. Generally, weed management practices such as herbicide use and mid-season tillage had the highest positive significant correlation with D' in August, while fertilizer input (organic and inorganic N, inorganic P) and rice coverage (with negative), and growth duration (with positive) had the highest correlation with these indices in October; and water depth and growth duration had the highest positive correlation with these indices in December. Soil humidity and burning had large correlation coefficients with D' and species richness in February.

WEED COMMUNITIES

The ordination results for the 224 quadrats on the two principal axes of the DCA, with the results for the six paddy groups overlaid. The correlations between the DCA scores and the site and vegetation parameters.

Table: Correlations between the DCA scores (axis 1 and axis 2) and the key variables on environmental conditions, vegetation and rice management for the combined analysis.

	Combined analysis	
	Axis 1	Axis 2
Physical parameters		
Distance from the main canal	0.248***	0.235***
Rice parameters		
Rice yield	–0.287**	–0.148*
Rice height	0.241**	0.143*
Rice coverage	–0.022	–0.202**
Weed parameters		
Weed height	0.117	–0.034
Weed coverage	–0.358***	–0.195**
Species richness	0.097	–0.178**
D'	0.380***	–0.006
Management input parameters		
Organic N	–0.331***	–0.150*
Inorganic N	–0.462***	–0.091
Inorganic P	–0.315***	–0.060
Herbicide	–0.153*	0.105
Weeding	0.171*	–0.080
Management parameters		
Percentage of DS/TP	0.848***	0.023
Number of days after planting	0.220**	0.570***
Rice growth duration	0.687***	0.064
Water environment parameters		
Water depth	0.317***	–0.144**
Soil moisture content	0.187**	–0.333***
Soil hardness	–0.045	0.135*
pH	–0.210**	–0.285***

Note:
Spearman's ranking correlations.
* $P < 0.05$.
** $P < 0.01$.
*** $P < 0.001$.

Axis 1 (eigenvalue = 0.779) was significantly positively correlated ($P < 0.001$) with the ratio of fields by direct-seeding to transplanting (1), number of days of rice growth (2), D' (4), water depth (7), and distance from the main irrigation canal (10), and it was significantly negatively correlated ($P < 0.001$) with inorganic N (3), integrated weed cover (5), organic N (6), and inorganic P (8), with the numbers in parentheses indicating ranks. Axis 2 (eigenvalue = 0.542) was significantly positively correlated ($P < 0.001$) with number of days after rice planting (1), and distance from the main irrigation canal (4), and it was significantly negatively correlated ($P < 0.001$) with soil moisture content (2) and pH (3).

Fig. the DCA results for the mean value for each paddy group in each of the surveyed months. Along axis 1, the six paddy groups were roughly in the following order from left to right: 1U, 1M, 1D, 7U, 7M, and 7D. 1U and 7D were clearly separated, whereas 1M, 1D, 7M, and 7U were closely grouped. Axis 2 of the DCA indicated the direction of seasonal changes in the vegetation community, and values along this axis generally increased from August to October to December to February. The changes along axis 2 were smaller for 1U than for the other five paddy groups. The changes for 1D, 7U, and 7M overlapped to a considerable extent.

DCA ordination was performed for the 52 weed species (i.e., excluding rare species from the total 76 species) in order from 1U to 7D along axis 1. 1U indicators had low scores (e.g., Ipomoea aquatica, P. spinescens, M. crenata, Eclipta prostrata, I. confusa), whereas 7D indicators (e.g., S. bispinosa, Ludwigia adscendens,A. sessilis) had high scores. Along axis 2, Melochia corchorifolia, Phyllanthus virgatus,Cyperus haspan, Nitella sp., and Lindernia angustifolia, which often occurred in August or October, had low scores, whereas C. procerus, Sphaeranthus indicus, Sacciolepis indica var. indica, and O. rufipogon, which appeared from December to February, had high scores.

UTILIZATION OF WEEDS BY FARMERS

As many as 13 weeds names were listed by farmers of KPIRA that they utilize. Some names referred to a single species (e.g., I. aquatica, Nymphaea nouchali, Monochoria vaginalis), whereas others referred to groups of species (e.g., some kinds of sedge or grass species). The amount of each weed collected from paddy fields and used for human consumption was generally small, whereas the abundant grass species were used as cattle forage. Farmers collected not only weeds from the paddy ecosystem, but also rice straw (as a forage), palm fruits, bamboo shoots, and trees (as firewood). To feed cattle, rice straw and weeds in the paddy ecosystem were the most widely utilized resources by farmers in both the rainy and dry seasons, followed by grazing in paddy fields after harvest.

DISCUSSION

In an irrigation rehabilitation area in Cambodia, 76 weed species were identified (in 224 quadrats totaling 896 m2). Previous surveys of paddy weeds in other tropical Asian countries showed similar numbers of species when excluding those growing on levees (41 species in Sri Lanka, Bambaradeniya et al., 2004; 87 species in Laos, Kosaka et al., 2006; 96 species in Thailand, Tomita et al., 2003a and Tomita et al., 2003b), whereas the inclusion of levee species markedly increased the numbers (89 species in Sri Lanka,Bambaradeniya et al., 2004; 184 species in Laos, Kosaka et al., 2006).

The Cambodian weed community in KPIRA shares more species with the rain-fed rice ecosystem in Laos, where both F. miliacea and Rotala indica were widespread, whereas only the former was widely distributed in rain-fed lowlands in northeastern Thailand (Tomita et al., 2003a and Tomita et al., 2003b). In the Vietnamese rice ecosystem, where the proportion of irrigated area was much larger than that in Cambodia, F. miliacea was not the most serious weed. Instead, Echinochloa crus-galliand Leptochloa chinensis were the species with more negative effects on rice production. Thus, the results indicate that the weed community in the Cambodian KPIRA had a greater similarity with rain-fed ecosystems than with irrigated ecosystems (i.e., Vietnam).

KPIRA had unique hydrological conditions in which irrigated rice had only been partly introduced. Some weed species common in rain-fed paddies in Cambodia were observed in KPIRA, such as Pentapetes phoenicea, Fuirena ciliaris, Lindernia crustacea, and Scirpus juncoides, as were species found in the floodplain of Lake Tonle Sap in Cambodia, including P. phoenicea, Aeschynomene aspera, and N. nouchali (Y. Araki, personal communication).

INTENSIFICATION AND WEED BIODIVERSITY

The study supported the hypothesis that a rich diversity of plants was available in KPIRA not only because of heterogeneous water conditions and crop management practices between upstream and downstream fields, but also because of low level of intensification by 2008 (i.e., low input of pesticides and inorganic fertilizers, extensive planting and weed control methods, mosaic field arrangement).

The presence of M. crenata, a perennial fern and a critically endangered species in the Japanese Red Data Book due to its sensitivity to agricultural chemical substances, indicated that agricultural input remained low in KPIRA. A negative relationship was found between the application rate of inorganic N fertilizer and Simpson's diversity index and species richness in October.

Upstream and D2-1 fields (1U, 1M), with higher organic and inorganic N and P fertilizer inputs and with relatively widespread transplanting, had a lower diversity index, whereas downstream and D2-7 fields (7D, 7M, 1D), with less fertilizer use, direct seeding, and mid-season tillage weed management, had a higher diversity index. Tomita et al. (2003a) reported richer weed species diversity by dry direct seeding compared with by transplanting in rain-fed lowlands in northeastern Thailand.

The paddy ecosystem in KPIRA consisted not only of rice fields and irrigation and drainage canals, but also levees, fallow fields, on-farm ponds, and small island-like mounded areas inside rice fields where trees grew (e.g., coconut) and provided places for farm workers to rest. Such a mosaic nature of landscape also helped to increase weed biodiversity.

This study supported the hypothesis of no trade-off between rice yield and weed diversity partly due to the relatively low level of intensification. However, greater extent of intensification such as an increase in application rate of inorganic N fertilizer would be demanded in future for KPIRA in order to increase rice yield and to meet with the Cambodian government's policy to enhance rice export (RGC, 2010). If policy and agricultural extension focused only on increasing rice yield, the use of much more herbicide and inorganic fertilizers would be encouraged, but this might dramatically reduce paddy weed diversity. Substantial numbers of plant species that grew in the paddy ecosystem of KPIRA were utilized by farmers for human consumption and forage for cattle and hence the merits of high weed diversity and weed provision ecosystem services deriving from the heterogeneity of rehabilitating irrigated rice ecosystem should be recognized. The strategy to increase rice yield further without reducing weed biodiversity could be possible for example by efficient application method of fertilizer N and improved management in direct seeding (Hayashi et al., 2007 and Ikeda et al., 2008).

The increased frequency of burning crop residues after harvesting (e.g., in 1M) caused a dramatic loss of weed biodiversity in February, when only limited species appeared (e.g.,M. crenata, S. grossus, and C. pulcherrimus). Burning could be an attractive residual management option for farmers to prepare for the subsequent planting as double rice cropping would further expand and cropping interval would become more intensified in KPIRA. However, burning had the negative aspects such as greenhouse gas emission, nutrient loss, diminished soil biota, and reduced total N and C in the topsoil (Gupta and Sahai, 2005 and Wassmann et al., 2009), as many governments in developing Asian countries made it illegal to burn crop residues. Farmers might lack an understanding of these negative aspects of burning for environment, as well as knowledge of available technology for in situ incorporation of the residues.

5

Weed Management in Organic Rice

The major impediment in the cultivation of rice is the heavy weed infestation, which compete with the crop to such an extent that the crop gets smoothered by the weeds. The weeds shared not only the plant nutrients but transpire a lot of valuable conserved water from the soil. The weeds also sometime serve as host for breeding and development of certain disease and pests. The delay in first weeding beyond 15-25 days after seeding sharply reduces the rice yield. Rice field weed species may conveniently be classified into broad leaved, grasses, sedges and others.

High fertility accompanied by high moisture provides a situation where the intensity & growth of weeds are high and in many cases may adversely affect the production potential of crops. Let us take into consideration the world's most important food crop 'Rice' that is the staple food of over half of the world population. The estimated yield losses in rice caused by weeds are given below:

Rice (transplanted) - 30-40% (Bhan(1994);

Rice (drilled) - 70-80% (Balasuberamaniyan & Palaniappan)

The weed problem in upland rice is more serious as compared to transplanted rice. The important weed species, which act as constraints in production potential of rice.

Weed management: Management of weeds is an important component of production techniques as elimination of weeds is expansive and hard to achieve. Presence of weeds is a constraint & their improper management further accentuates the effect. In the past two decades work has been done on non-chemical management techniques and environmentally safe alternatives to herbicides for weed control. Such ecofriendly techniques for weed control in rice fields are given below:

Preventive measures:

1. Always use well rotten and decomposed organic manure.
2. Avoid feeding of grain screenings or hay containing weed seeds without destroying their viability by grinding or cooking, otherwise weed will spread from dung or manure.

3. Clean all the implements & machinery properly after their use in infested areas & before using in clean areas.
4. Keep irrigation and drainage channels free from weeds.
5. Watch seedlings in nursery carefully so that they do not get mixed with weed seedlings & get carried to the fields.
6. Use of weed free seeds/seedlings for sowing/transplanting from a reliable source. It is a preventive method against introduction of weeds.

Under these practices, use of clean seeds for sowing and of weed free seedlings for transplanting should be practised. New exotic introduction of weed species perhaps (seed/seedling contamination) is the one of the reason and if care is taken timely, it can be managed easily.

summer ploughing: Ploughing in summer exposes the underground parts like rhizomes and tubers of perennial and obnoxious weeds to scorching summer sun and kills them. It also helps in improving the soil physical conditions. Conventional tillage that includes 2-3 ploughings followed by harrowing decreases the weed problem in upland rice.

Field preparation: Better land preparation (2 ploughing at 15 DBS and 2 at sowing), timely sowing (last week of June), Optimum management of nutrients and an additional hand weeding markedly decreased the infestation of all category of weeds in rainfed upland rice. In lowland rice, puddling operation incorporates all the weeds in the soil, which would decompose in course of time & that is the reason for reduced weed infestation in transplanted rice as compared to direct seeded rice.

Planting method: Sowing should be done 2-3 days after irrigation in upland rice. Weeds already present in soil start germinating with 2-3 days. Sowing operation with seed drill removes some of germinating weeds as blade harrow is run to cover the seeds. During this process, all the surface soil to a depth of 2-3 cm is disturbed,thereby uprooting the germinating seeds. In addition due to loosening of surface soil, it dries up quickly & does not allow weed seeds to germinate until subsequent rain or irrigation. For this reason weed population is high when irrigation is given after sowing or rain is received after sowing.

Transplanting is another operation that reduces weed population since the crop has an additional advantage due to its age as a result covering the ground early. The clean field preparation for transplanting helps in reducing weed germination. Weed population is more in direct seeded rice as compared to transplanted rice.

Bed planting or ridge planting systems also helps in reducing weed population in rice. In these systems, we can easily control the weeds in the furrows between the rows of rice crop by mechanical methods like manual weeding, blind tillage, flooding, burning, mulching, solarization etc.

The main processes involved are germination and emerging of seedlings from the seed or establishment of seedling after transplanting. The best method is bed planting for both direct seeded and transplanted rice. It was found more economical in terms of saving of water, weed management and finally the yield. Direction of beds did not differ significantly. Crop-weed competition studies have shown that biomass and reproductive potential of weeds is significantly reduced if the light competition by the crop is improved (Baumann D.T.,Bastiaans L., Kropff M.T.,1999). Thus weeds remained under check on beds with two rows. Furrow is irrigated + mulched with straw. The total population remained the same. Direct seeding in time on flat method was also found good if weed free conditions are maintained upto four weeks after germination of crop.

Relay planting at different places showed minimum weed population (Broadcasting of rice seeds in standing wheat crop with last irrigation applied).

INTERCROPPING

Mixing of legumes will cover the ground quickly and many workers have observed smoothering effect on weeds especially in direct seeded rice. In Operational Research Project, the rice + sorghum, rice + Hibiscus spp or rice + teosinte seeding geometry found to be the best because of uneven distribution of rainfall and accumulation of flood water in plain zone of Haryana state of India.

One hand weeding 3 weeks after sowing recorded superiority in managing the weeds.

Plant population: Plant population & row arrangement also affect the weed growth. Higher plant population & narrow row spacing can put pressure on the availability of space for weed growth. Closer planting of crops suppresses germination and growth of weeds due to allelopathy and competition for growth factors. A narrow (15cm) spacing was found superior to wide (30-45cm) spacing in minimizing weed competition & increasing productive tillers & yield in upland rice. Research showed that increasing the seed rate from 40 to 100kg/ha for direct seeded rice decreased weed weight from 52 to 188g/m2 and increased rice yield from 2.78 to 3.45t/ha. Higher of plant population in transplanted rice had lower dry weights of weeds.

Blind tillage/Stale seedbed: This method is extensively used to combat weeds in rice in early stages before the sowing or just after sowing but before the crop emergence. In this method, weeds are allowed to germinate and then their above ground parts are destroyed by using various types of blade harrows as soon as the weed appears.

Hand pulling and hand hoeing: Despite major advantages in chemical control, hand removal of weeds still remains to be the most practical method of weed control in many developing countries. Although back breaking &

laborious, hand weeding is quite effective if employed at the right time. Two aspects are important in hand weeding: the no. of hand weeding to be done & interval between two hand weeding. The number of hand weeding depends on crop growth, weed growth & critical period of crop-weed competition.

Hand hoeing is done in upland crop, the entire surface soil is dig to shallow depth with hand hoes, weeds are uprooted & removed. It also improves soil physical condition.2-3 manual weeding at 25 and 45 DAS in direct seeded rice and at 15 and 30 DAT in transplanted rice is the best and most effective way of weed management.

Post plant tillage: Rice rotary weeder is a specific tool for weeding in rice fields. Single or double row weeders are also available. They are worked manually with back and forth movement in between rice rows in wetlands. This rotary weeder is less strenuous than hand weeding. At the time of weeding, it is essential to retain 1.5-2.0 cm depth of water. With the single row rice weeder, one labor can cover about 0.5 ha/day. 2-3 weeding with paddy weeder at 15 and 30 DAT in transplanted rice is the best and most effective way of weed management.

MECHANICAL METHODS

Under this, different tools & implements have been tried. Irrigate the field before planting and allow the first flesh of weeds to germinate which are controlled by the bullock drawn disc harrow and then seeding the rice. In North India, in last week of May, irrigate the field to a depth of 40mm of water and after 10 days, disc harrowing (bullock drawn) is done to prepare the field. Then sowing of rice is to be completed within two days. Consequently, threshold value is achieved.

Various types of hoe & weeder have been tried in the standing crop of rice. which worked manually with back & forth movement in between rice rows. Just like a wheel hand hoe and one person can weed 0.5 ha/day (8 hours work). Work is in progress for sensor technique attached to mechanical weeder, which identify and suppress the weeds.

Recently zero-tillage technique has been successfully demonstrated at farmers' field. Data collected indicate that weed weight in dry season is less under zero-tillage as compared to conventional tillage seeding method. Similar trend was followed to weed counts both for grassy & broad-leaved weed species. It is more economical method in terms of time, energy and money (unpublished data).

Solarization: It involves covering the soil surface with polythene sheets to increase the soil temperature, which would be lethal to weed seeds. A mulching period of 2-6 weeks with clear polythene sheets has been reported to give effective control of many annual weeds. Irrigation prior to solarization has a complementary effect, as moisture-imbibed seeds are more sensitive to heat than dry seeds. The limitation however is that normally weed seeds up to about 5 cm depth are only affected implying that deep prepatory cultivation would nullify the effect. Although this technique is limited by the cost of treatment, it may be made use of in controlling weeds in nursery areas. Mulching by polythene sheets for 32 days decreased emergence of Dactylotenium aegypticum and Cyperus rotundus by more than 90% and the main solarization is restricted to the 0-5 cm layer of soil.

MULCHING & MOWING

Different types of mulch like straw, hay, saw dust, paper, plastic film etc. when applied to soil surface, do not allow weeds to germinate or to grow as light does not reach to the soil.

This has been practiced for obnoxious weed as well. It has been demonstrated that harvesting the weeds with sickle between the rows and placing the harvested material like mulch there and this gave good management of weeds.

Moisture loss and weeds remained under control with no investment. In bed-planting method of rice seeding, the furrow is covered with straw mulch gave sufficient control of weeds. Spreading of polythene sheets for 32 days in furrow checked the emergence of Trianthema, Cyperus & Dactylotenium spp under direct seeded upland rice.

Critical period of crop-weed competition: This period indicate the critical crop growth stage or the period during which the field must be kept weed free. It is that shortest time span during the crop growth when weeding results in highest economic return. Generally early one-third duration of the crop period should be maintained weed free. The critical period also differs with the cultivars used.

For examples, 20 days of weed free growth appears best in short-statured variety of rice. For an intermediate-statured variety, the weed free period should be extended to the first 30 days after planting of rice. Generally weed competition is less under flooded condition where rice is transplanted as compared to upland situation.

Varieties: Short statured, erect leaved varieties permit more light compared to tall & leafy traditional varieties. Some of the crops grow quickly & produce canopy early resulting in shading and thus suppress weed growth. In order to control the weeds like Echinochloa colonum which one morphologically similar to rice, a pigmented variety 'kalashri' (R260-292), grown in Orissa state, India, may be useful in distinguishing weeds from rice in early growth stage.

New introductions do have impact on weed ecology in a region. Selection of crop spp or varieties should be tested and tried in small protective areas. Genotypes introduced from IRRI have changed weed flora in Northwestern plain zone of India. Consequently today Phalaris spp has become most common resistant weed in rice-wheat system of this zone due to spread of short duration & high input responsive cultivars. Whereas, this weed is not a problem with traditional varieties. Experiments at cultivator's field (ECF) indicate that genotype in relation to environment do have effect on weed flora of that field. 'Kalinga-3', a pigmented tall rice variety is superior to smoother the weed E. colonum.

Crop rotation: In monoculture, a number of weed spp persist and expand rapidly and increases the possibility of occurring of certain weed spp or group of spp.

Crop rotation helps to interrupt the life cycle of weeds and thereby can eliminate or at least reduce difficult weed problems. Weeds like Cyperus rotundus (motha) can be controlled effectively by including lowland rice in crop rotation in place of upland rice. Add the smoothering crops in rotation like fodder grasses and majority of pulses controls the weeds very effectively. In continuously transplanted rice fields under irrigated conditions, Scripus maritimus persists. When upland crops are rotated with rice, the population of this weed has been reduced drastically even without adoption of weed control measures. Similarly,

Striga spp, a difficult weed in continuous flooded low land fields was effectively controlled when rice was rotated with an upland crop.

The old common practice, which farmers used to follow, is the rotation and it needs no explanation. Every farmer knows this technique round the globe.

Under organic farming it serve as the best tool for carrying out the various activities like change of soil micro flora, release of toxin, which in turn improve soil fertility.

The weed seed banks also affected due to rotation and consequently weed infestation found at lowest count. Many times it was recorded interaction between site condition and cropping system. Sensitivity analyses clearly indicate count below the threshold values for weeds in direct seeded rice. Under puddled condition (which is not acceptable in general) scenario changed. Whereas, flooded rice cultivation have its limitation for varied rotation to be followed.

Growing of inter crops not only produce higher return but also help in reducing of weed infestation in row crop. Growing of cowpea in between the rows of upland rice resulted in an energy saving up to 315MJ/hacompared to recommended practice and specific energy requirement is 42.0MJ/ha than hand weeding

Water management: Depending upon the method of irrigation, weed infestation may be increased or decreased. Frequent irrigation or rain during initial stages of crop growth reduces several flushes of weeds. In lowland rice, where standing water is present most of the time, germination of weeds is less, which is mainly due to puddling & impounding of water.. Weed population and type of seeds changes with the depth of water in rice fields. Continuous submergence with 5cm water results in reducing weed population whereas under upland situation, weed population and weed dry matter is very high.

Effect of irrigation levels on weed growth:

Irrigation level	No. of weeds/m2			Weed dry matter(kg/ ha)
	Grasses	BLW	Sedges	
Submergence-5cm	5	0	0	6.83
Submergence to saturation	12	5	0	4.36
FC to 20% DASM	382	11	51	73.43
FC to 50% DASM	494	16	9	72.15

DASM- Depletion of available soil moisture,FC- Field Capacity,BLW- Broad-leaved weeds

Under submerged condition, tubers of Cyperus rotundus killed due to lack of oxygen. Flooding upto 10-20 cm early in the season reduced the infestation of many weeds including Echinochloa crusgalli. Drainage or alternate wetting & drying induce germination and increases weed problem in rice.

Under irrigated conditions, planting crop-seeds in the moisture zone in an otherwise dry seedbed and delaying the first irrigation reduces the weed infestation quite substantially.

Biological controls: It involves deliberate introduction & establishment of natural enemies in areas where they did not previously occur. Good control of a variety of aquatic weeds by Chinese grass carp (Ctenopharyangdon idella) has received worldwide attention. In Japan, there is limited use of tadpole shrimp (Triopus spp) to control weeds in flooded rice fields. At the IRRI, the cover effect of Azolla spp is being evaluated as a means of weed control in rice.

Several species of herbivorous fish feed on aquatic weeds. The most important fish are Tilapia melanopleura, T. zilli, T. nilotica, and Puniux gonianatus. In Japan, farmers also used some species of ducks for weed control in natural rice farming.

BIOHERBICIDES:

Rust fungi have been tried for control of Cyperus spp.

Synthetic derivatives if naturally occurring compounds as herbicides:

Methoxyphenone is the herbicide developed as a synthetic analog of the microbial toxin, anisomycin by employing bio-rational chemical synthesis of an herbicide based on a natural chemistry. In Japan, this herbicide is marketed as a selective herbicide for the control of banyard grass (Echinochloa crusgalli) in rice is easily degraded in soil.

Natural herbicides: Solutions containing rice hull extract inhibits the germination of barnyard grass (Echinochloa crusgalli). Japanese farmers uses rice bran (200 g/m2) for weed control.

Many crop residues including straw from wheat, barley, rye and sorghum contain allelopathic compounds. Volatilization, leaching, root exudation, and decomposition of plant residues release these compounds. Corn gluten meal (CG) is a by-product of corn wet milling process and has been patented as a natural pre-emergence herbicide.

6

Utilization of some Weeds as Medicine

Since the dawn of human civilization, men have used plants as a source of medicine, because they were easily available in the immediate environment. The most effective plants from them were selected and now they are part of ethno medical traditions. Weed species interfere with our endeavors, such as agriculture or animal farming, recreational pursuits, including gardening, transport, bush walking and water sports etc. Globally, the utilization of weeds has been patchy over the past few decades. Nevertheless, there is a renewed interest in focusing on utilization of weeds in productive ways, so that people may benefit from an aspect that has been largely ignored (Chandrasena, 2007). 'Utilization' has been recognized as an effective means of weed management.

Weeds have been used for long time as sources of food, fiber, dye, medicines, etc. Unlike other crop plants, weed plants are less vulnerable to disease and insect attack. Plants provide various kind of drugs and medicines and our dependence on medicinal plants has no way been minimized by the use of modern system of synthetic drugs whose use are not without side effects. In India, there are more than 7000 species, which have been identified as medicinal plants. The indigenous system of medicine practiced in India is based mainly on the use of plants. Charaka Samhita (1000 BC-100 AD) has recorded 2000 vegetable remedies.

There are about 120 weed species in India used as medicinal purpose (Naidu et al., 2005). Ancient medicine was not solely based on empiricism and this is evident from the fact that some medicinal plants which were used in ancient times still have their place in modern therapy. Weeds are highly valued in traditional medicine systems and have been used by indigenous communities for curing different ailments for thousands of years.

Most of weeds have been known to posses therapeutic properties and the pertinent traditional knowledge was transferred orally through generations. Use of crop field weeds as folk medicines in Bankura district have been reported earlier (Mukhopadhya and Duary, 1995; Mukhopadhya et al., 1995; Mukhopadhya and Duary, 1997). It is clear that new pharmaceuticals are like to be found in colonizing plants, and as Stepp (2004) suggested, weeds need to be given more attention as potential sources of phytomedicines. This is important because according to an estimate of WHO, approximately 80% people of developing countries rely chiefly on traditional medicine for primary healthcare (Ghosh, 2008). Plants have been used as a source of medicine for living being from ancient period of time. Documentation of traditional knowledge of ethno medicinal use of plants has been considered as a high priority to support the discoveries of drugs benefiting mankind. The tribal populations, who have been the primary inhabitants of natural habitats, hold tremendous amount of traditional knowledge on the use of various biotic resources, which may have greater importance to the ongoing research and discoveries in the field.

It is well acknowledged in literature that their age old practices of using plants to cure numerous ailments have paved the way to further the discovery of many life saving drugs. This represents over 550 medicinal plants, which may offer incredible scope for the development of pharmaceutical sector as potential commercial hub, boosting economy of the state. Ethno-medicinal explorations and simultaneous prioritization of pharmaceutically important plant species for conservation through ex-situ cultivation have been identified as vital aspects for the drug industrial development. Plants have been used in the traditional healthcare system from time immemorial, particularly among the tribal communities. Numerous wild and cultivated plants play a vital role in their culture, customs, traditional healthcare system, rituals, etc., and this interrelationship has evolved over generations of experience and practice.

Ayurveda, which is one of the oldest systems of traditional healthcare system and yet living in traditions practiced widely in India, Sri Lanka and other countries has a sound philosophical and sound basis. Atharvaveda (around 1200 BC), Charak Samhita and Shusrut Samhita (1000-500 BC) are the main classics that gives a detailed description of over 700 herbs. Herbal medicines are becoming popular worldwide due to its growing recognition of natural products being cheaper and without any side effects. Demands for medicinal plants are increasing in both developing and developed countries. Out of the 20,000 medicinal plants listed by the WHO globally India's contribution is 15-20%. In India, about 2,000 drugs used are of plant origin.

India contains over 5% of the worlds' diversity though it covers only 2% of the earth's surface but it is also one of the biodiversity hotspots of the richest and highly endangered eco-regions of the world. At present, there is a worldwide movement for assessing the plant resources and researches for new plants which are of medicinal and economical value and importance. Researchers are focusing mainly on ethnobotanical and ethno medicinal investigation to fulfill the increasing demand of herbal products. The traditional knowledge of herbal medicine is much enriched here in the district due to its diversified plant wealth and this valuable knowledge which is still surviving in the tribal culture has to be documented immediately before it gets lost forever. With this perspective present investigation has been carried out to provide the significant information regarding traditional uses of weeds as folk medicine by the local people of Birbhum district, West Bengal which are the new addition to the district inventory of ethnomedicine.

MATERIALS AND METHODS

The present study deals with the indigenous knowledge related with ethnomedical uses of weed plants used by the local people of Birbhum district, West Bengal. The district Birbhum is one of the smallest district of the state West Bengal, India that is full of natural resources. Birbhum is bounded on the north and west by Santhal Paraganas, on the east by the districts of Murshidabad and Burdwan and on the South by Burdwan district, from which it is separated by the river Ajay. The district extends over an area of 4545 sq.km and it is situated between 87010/ & 8802/ east longitude and between 23033/ & 24035/ North latitude. The temperature varies from 110C (in January) to 42.90C (in May).

The annual average rainfall is 1098mm. Total population of the district Birbhum is 3015422 which is 2.45% of the state population. In the district Birbhum, the tribal population is 203127 contributing 6.73% of the total population of the district. The major tribal groups of the district are Santal, Kora and Oraon. The rural people mostly depend on cultivation. Prior to the field visits, extensive literature survey was carried out on the previous ethnomedicinal and floral reports on the district. Rural areas were visited during summer, monsoon and winter to avail most of the plants in their conditions. During the visits, the informants were chosen on the basis of structured questionnaire.

The methodology was adopted as described by Jain (1999), Chadwick and Marsh (1994). Structured questionnaires, interviews and participatory observations were used to elicit information from the resource persons using standard methods (Martin, 1995). The data was recorded in a data sheet with the names of the plants, families, local names, parts used, ethnomedicinal uses. Informants were selected on the basis of their ability to identify a particular plant in situ and their basic knowledge of ethnomedicine. Local herbal medical practitioners (folk doctors) and elderly people were preferred during the interviews.

Generally the two types of interviews were taken, firstly of individuals and secondly of groups. Of individuals, persons were selected at random on the way or entering a hut finding out knowledgeable individuals from the village or also the Headman. In group interviews more than one individual were approached, our purpose explained and interviews taken. They were requested to collect specimens of the plants they knew or to show the plant species on site. The collected plant species have been carefully identified with the help of different Floras and standard literature (Datta and Banerjee, 1978; Jain, 1987; Jain, 1991; Sanyal, 1994; Maheswari, 2000; Tribedi and Sharma, 2004). The plant specimens have been preserved as herbarium specimen following conventional techniques (Jain and Rao, 1977).

RESULTS AND DISCUSSION

Locally available weed plants are used by the peoples for their household remedies and various purposes. The data has been verified from the ethnic people of different tribal areas. Information for treating a particular ailment from different informants certainly reflects the accuracy and authenticity of the folk drugs employed.

The major weed flora associated with transplanted rice in the district are Alternanthera philoxeroides, Alternanthera sessilis, Ammania baccifera, Commelina nudiflora, Cyanotis axillaris, Lindernia ciliata, Lindernia crustacea, Ludwigia parviflora, Marsilea quadrifolia Ammania multiflora, Jussia repens, Sphenoclea zeylanica, Croton bonplandianum, Portulaca oleraceae, Leucas aspera, Asteracantha longifolia, Bacopa monnieri among broadleaved, Cynodon dactylon, Dactyloctenium aegyptium, Digitaria sanguinalis, Echinochloa colonum, Echinochloa crusgalli, Echinochloa glabrescens, Paspalum distichum among grasses, Cyperus difformis, Cyperus iria, Cyperus compressus, Fimbristylis miliacea and Scirpus articulatus among sedges and Chara zeylanica, as algal weeds.

Weed flora in rabi crops are Anagallis arvensis, Celosia argentea, Chenopodium album, Croton bonplandianum, Drymaria cordata, Eclipta alba, Gnaphalium indicum, Gnaphalium pensylvanicum, Gomphrena celosioides, Gnaphalium purpureum, Oxalis latifolia, Physalis minima, Polygonum plebeium, Solanum nigram, Spilanthes acmella Melilotus alba, Melilotus indica, Physalis minima, among broad leaved; Cynodon dactylon, Digitaria sanguinalis, Echinochloa colonum among grasses and Cyperus rotundus as sedges.

Major weeds associated with sugarcane are Argemone mexicana, Ageratum conyzoides, Blumea lacera, Commelina nudiflora, Convolvulus arvensis, Eclipta prostrata, Eclipta alba, Lindernia crustacea, Ludwigia parviflora, Gomphrena celosioides among broad leaved; Cynodon dactylon, Dactyloctenium aegyptium, Digitaria sanguinalis, Echinochloa colonum among grasses and Cyperus rotundus, Cyperus compressus among sedges.

The chilli crop is infested with Gnaphalium indicum, Gnaphalium pensylvanicum, Gnaphalium purpureum, Phyllanthus fraternus, Physalis minima, Solanum nigrum, Tridax procumbens, Vernonia cineria. Weeds associated with summer vegetables are Alternanthera sessilis, Amaranthus viridis, Croton bonplandianum, Euophobia hirta, Trianthema portulacastrum among broad leaved weeds; Cynodon dactylon, Digitaria sanguinalis, Echinochloa colonum as grasses and Cyperus rotundus as sedge (Anonymous 2010).

In our investigation we found that tribal people of Birbhum district use some of these weed plants in different ways which is devoid of Ayurvedic and Unani medicinal system. These are as follows:

Table. Weeds used as medicine by the local people of Birbhum district

SL.No	Botanical name	Family	Tribal/Local name	Parts used	Ailments	Mode of administration
1	*Echinochloa colona*	Poaceae	Shyama ghas/Lakshmi ghas	seed	digestive stimulant tonic for the liver	Cooked or eaten raw with rice
2	*Argemone mexicana*	Papaveraceae	Shialkanta	Whole plant, Leaves, seeds, roots flowers, fruits. yellow juice, latex	Analgesic; Antispasmodic; Antitussive; Demulcent; Emetic; Expectorant; Hallucinogenic; Purgative; Sedative; Skin; Warts. Impotence	The root has been used in the treatment of chronic skin diseases. The flowers areexpectorant and have been used in the treatment of coughs. The seed has also been used as an antidote to snake poisoning. The oil from the seed is purgative. It has been used in the treatment of skin problems.
3	*Cyperus rotundus*	Cyperaceae	Mutho ghas	Root	Digestion	To make paste with small amount of salt and should be taken daily after major meal

4	*Alternanth era sessilis*	Amaranthac eae	Sinche sak/ Senchi sak	Whole plant	burning sensation, diarrhea, skin disease, dyspepsia, hemorroids, liver and spleen diseases and fever.	Plants are made into paste and paste is administer ed as poultice on affected body part.
5	*Ammania baccifera*	Lythraceae	Dadma ri	Whole plant	burning sensation	Paste of whole plant
6	*Commelina nudiflora*	Commelinac eae	Kansira	Whole plant	skin disease	Juice of whole plant
7	*Marsilea quadrifolia*	Marsiliaceae	Susoni sak/Sus ni sak	Whole plant	cough, bronchitis, diabetes, psychiatric diseases, eye diseases, diarrhea and skin diseases	Used as vegetables and juice of leaves is applied on affected area.
8	*Echinochlo a crusgalli*	Poaceae	Moland a, Dotala ghas	Whole plant	spleen and in checking haemorrhag e	Take juice daily
9	*Cynodon dactylon*	Poaceae	Durba, Dubba ghas, Duburi ghas	whole plant Leaf	congestive heart failure, diarrhea eye tonic, prevent conjunctiviti s	Drink juice daily
10	*Chenopodi um album*	Chenopodiac eae	Bethu, Beto sak	Leaves	Appetizer, dysentery, digestive	Leaf as vegetables
11	*Polygonu m plabeium*	polygonacea e	Maskati , Chikni sak	Whole plant	bleeding, dysentery and haemorrhoi ds	Fresh juice is applied in affected area and leaves are used as vegetables
12	*Taraxacum*	Asteraceae	Dandeli	Whole	digestive	To make

13	*Ludwigia parviflora*	Onagraceae	Bon Labang a, Saga ghas	Root	Cancer	Drink Juice daily
14	*Croton bonplandia num*	Euphorbiace ae	Ban tulsi, Bhabri	Stem Leaves	Clotting of blood, curing of wounds	Leaves juice and latex are applied
15	*Centella asiatica*	Umbelliferea e	Thalkur i, Thanku ni	Twigs, Leaves	Blood dysentery, Appetizer	Fresh leaves and twigs in empty stomach for 5 – 7 days cure blood dysentery
16	*Eclipta alba*	Asteraceae	Kesut, Keshuk ti	Leaves Whole plant	Cooling effect on brain, Skin diseases	Fresh leaves are applied with sesame oil to cure baldness/el ephantiasis & headache and juice of whole plant is applied on affected area on skin.
17	*Portulaca oleracea*	Portulacacea e	Nunia sak, Luniya sak	Whole plant	Cooling in stomach, skin, dysentery	Whole plant juice boiled against dysentery and prickly heat in stomach
18	*Leucas aspera*	Lamiaceae	Drone, Halkus a	Leaves	Skin diseases, cough and cold	Fresh leaves juice applied in skin eruption, allergic swelling

There are numerous uses of weed species as herbal medicines to cure a host of body ailments and diseases in this region. Now, traditional knowledge regarding the use of medicinal plants has been threatened in its existence and is gradually being lost from the traditional society due to erosion of its culture.

It is now high time for us to document the herbal traditional knowledge before it gets lost from the tribal society forever and simultaneously to conserve these medicinal plant resources also. The data provided in this study will finally be helpful to prepare the district as well as state inventory on folk medicine. This traditional herbal knowledge of the studied areas needs proper documentation otherwise it will be lost from its folk society forever. The information documented in this investigation, will further be validated through phytochemical, pharmacological and clinical studies.

The various folk medicinal uses of plants recorded here in this investigation need further scientific studies for their therapeutic validation. However, more research is needed to verify the active chemicals and how herbal medicines, based on weeds, can cure human diseases. Thus, investigating therapeutically or allelopathically active compounds in colonizing plants presents a scientific challenge. Elucidating the chemistry of these bio-active compounds will led to identifying opportunities for future development of medicines.

CONCLUSION

Weeds are clearly highly successful plants owing to their special characteristics that confer superior colonizing ability and competitiveness. These attributes can be useful in many situations, such as in repairing damaged ecosystems. Weed species which are used for medicinal plants will receive more attention because 80% of the world population continues to rely mainly on traditional medicines for their health care. Studies on verification of chemical component in specific medicinal weeds will give a clue of synthesizing a new medicine.

WEEDS IN AGRO-ECOSYSTEMS: AS A SOURCE OF MEDICINES FOR HUMAN HEALTH CARE

Plants have been, and still are, a rich source of many natural products. In India, most of which have been extensively used for traditional human health care systems viz. Ayurveda, Unani and Siddha. The vast majority of people in the world takes care of themselves and uses healing plants that have been used for hundreds of generations (1-4).

The most of the plants used by the rural communities have biologically active compounds that have been shown by generations to be effective against specific disorders. The global demand for herbal medicine is not only large, but also growing (5). In 1991, almost half of the best selling pharmaceuticals were either natural products or their derivatives (6). The market for Ayurvedic medicines is estimated to be expanding at 20 per cent annually in India (7). Only 15 per cent of pharmaceutical drugs are consumed in developing countries, and relatively more affluent people take a large proportion of even this small percentage (8). In some rural areas the collection of medicinal plants for the Indian market is a flourishing business. More than 50 per cent of households in the northern part of central Nepal and about 25 per cent in the middle part of the same region are involved in collecting medicinal plants for sale, the materials being traded on to wholesale markets in Delhi (7). Every year, the human knowledge about the distribution, ecology, methods of extracting the useful properties of plants and methods of management is declining rapidly. A continuation of the loss of local cultural diversity has underway for hundreds of years (9).

Forest is often considered to be the most promising habitat for source of this search due to high biodiversity and endemism (10). Many scientists have combined this assumption with an ethnobotanical approach to natural medicinal plants discovery in order to maximize the successful development of pharmaceutical products (11-14). Ethnobotanical surveys help the suitable source of information regarding useful plants and process of domestication which is a major evolutionary force bringing about different forms of plants through human selection (15). Now a day, collection of medicinal plants from forest is very difficult due to government's forest policy. Therefore, this focus on forests overlooks the fact that disturbed ecosystems are preferred habitats for medicinal plant procurement by many traditional peoples (16,17). Among the conservative estimate of 250,000 flowering plants in the world (18), more than 8000 species are weeds (19).

The weeds grow along with the crop plants (agro-ecosystems) and are regarded as nuisance for crops. But are the raw materials to the pharmaceutical industries as they yield chemicals used in formulation of various drugs, Vaidyas for preparing herbal formulations and an important source of medicines for indigenous peoples (21, 22). There are a number of reasons that the rural communities use weeds as medicine found in nearby areas (23). There is some evidence that the plants lose their effectiveness over time, must be used when freshly picked and effective when they grown in disturbed areas. Many weeds contain chemical compounds which are biologically active and potentially useful for medical science.

There is also good biochemical evidence that supports the hypothesis that plants in disturbed areas are likely to have more chemicals in them for defense. Today intensive agricultural practices and environmental degradation of habitats in many agro-ecosystems could have an impact on availability of ethnobotanically important plants. This may result in a conflict of interest regarding plant species that have a value to some communities but are regarded as undesirable by others. Further the introduction of large number of ruminants has resulted in overgrazing and reduction of vegetation in surrounding ecosystems.

The study of medicinally important weeds has not been realized as fully as other traditional communities elsewhere such as wild plants in forest ecosystems which often exclude weed species (13,14,24,25). In view of the rapid loss of diversity of plants, natural habitats, traditional community life, cultural diversity and knowledge of medicinal plants, documentation of medicinally important weeds is an urgent matter. Secondly, search for new medicines with low cost, more potential and without adverse side effect is needed to solve the major health problems. These efforts are in line with the global convention on biological diversity, 21st agenda and the biodiversity strategy of Tamilnadu Government (26,27). It helps to recognition in to popularization of economic importance of plants, upgrading herbal medical practices, conservation of indigenous knowledge and medicinally important plants.

MATERIALS AND METHODS

Study area

A vegetation survey was undertaken to determine the role that weedy plant species currently play as a source plants for traditional medicines during June 2004 to May 2006. The study area was north eastern coastal agroecosystem of Tamilnadu, India comprising Cuddalore, Villupuram, Kanchipuram and Thiruvalluar districts comes under the North Eastern and Cauvery Deltoic climatic zone. Geographically it lies between 11022' to 13028' N latitude and 79045' to 80020' E longitude and the altitude varies from 3 to 27 m amsl. The sampling was done up to 10 km inland from the coast and covers an area of 1800 sq km. The area received an annual mean rainfall of 1120 mm and a mean temperature of 31o C. The minor growing season starts from August – October and the major growing season is November – February, followed by a long dry season from March – July. The aborigines of the area are mostly farmers and daily wage groups.

A species was considered to be a weed if it was included in the standard reference for weeds worldwide based on a global literature search (19, 28-31). A weed that corresponds to those species included in this study is "a plant... if, in any specified geographical area, its populations grow entirely or predominantly in situations markedly disturbed by men, without, of course, being a deliberately cultivated plant" (32). All weed species were collected in the field by means of field interview (33) with semi structured questionnaires. Informants were asked to guide as to the places where these weeds grew or to bring the drug they use. Each interviewee was shown plant specimens collected and the medicinal property of each species was accepted as valid if at least twenty five per cent informants had a similar opinion. Additional discussions were conducted with the traditional healers including herbalists and diviners.

A sample of each medicinally important weed identified was preserved systematically (34) in the herbarium of Agronomy Department, Annamalai University, Annamalainager, Tamilnadu for their further reference. Finally additional information's regarding plant uses as medicine was noted and confirmed with the help of available literatures (35-45).

RESULTS AND DISCUSSION

The present investigation comprises 88 species of ethno medicinally important weedy plants distributed in 81 genera belonging to 43 families. Amaranthaceae was the most dominant family with 8 species, followed by Euphorbiaceae (7 species); Fabaceae, Malvaceae and Solanaceae (5 species); Asclepiadaceae, Compositae and Convolvulaceae (4 species); Capparidaceae, Molluginaceae and Poaceae (3 species); Acanthaceae, Boraginaceae, Cucurbitaceae, Pedaliaceae and Rubiaceae (2 species). Alternanthera sessilis, Centella asiatica, Commelina benghalensis, Cynodon dactylon, Eclipta alba, Marsilea quadrifoliata, Oxalis corniculata, Phylanthus niruri, Portulaca oleracea, Solanum nigrum, Solanum trilobatum, Trianthema portulacastrum and Tridax procumbens are the most commonly used medicinally important plants in the inhabitants of north eastern coastal Tamilnadu. This may be connected the fact that the popularity of the ailments that they are used to treating.

The direct use of popular medicinal plants as ailment is now very low for many inhabitants. They don't grow medicinally important plants in their gardens and collect these from their surrounding environments rather than buy, because these plants are used as an unexpected ailment. The weeds in agro-ecosystems are mostly annuals, they do not grow in the expected season and the people do not have the expertise or enough knowledge to the importance of these plants. The knowledge of these plants is passed from one generation to another verbally and through experience. Now the younger generations are not interested in agricultural activities and do not popular with traditional practices. The use of medicinally important plants is at a very low level due to lack of remunerative prices and market linkages. These are the factors encountered for lack of popularity as a raw material in the traditional human health care system.

Common health problems in the sites of the study area were external problems such as burns, cuts and wounds, cough, fever, headache, poison bites and skin diseases and the largest number of medicinally important weeds was used to treat these troubles. Common ailments such as headaches or coughs are considered to be diseases with natural causes and hence their symptoms are treated at the household level (46-51). In the present study nine remedies (Abrus precatorius, Acalypha indica, Aerva lanata, Cardiospermum halicacabum, Clitoria ternetea, Leucas aspera, Phyllanthus maderaspatensis, Portulaca oleracea and Vernonia cinerea) were used to get relief from headache. The inhabitants used Tridax procumbens as a major herb to cure cuts and wounds. Andrographis paniculata, Coccinia cordifolia, Portulaca oleracea and Solanum trilobatum are used to treat diabetes by the local traditional healers of this location. Averva lanta, Biophytum sensitivum, Phylanthus niruri and Tribulus terrestris are the recorded other plants used to cure diabetes. Many traditional plant treatments for diabetes are used throughout the world and there is an increasing demand by patients to use the natural products with anti diabetic activity (52).

In this locality "Jaundice" (Yellow Fewer) is considered as very serious disease. The most of the inhabitants and local traditional practitioners used Phyllanthus nirurai and or its combinations as an ailment to cure jaundice rather than other pharmaceuticals. But the literatures show that more than ten species of locally available plants (Andrographis paniculata, Argemone mexicana, Boerhaavia diffusa, Eclipta alba, Hygrophila auriculata, Imperata cylindrica, Leucas aspera, Mimosa pudica, Phyllanthus maderaspatensis and Solanum nigrum) are used traditionally to cure jaundice (14,20,41,53). The plants such as Abrus precatorius, Anisomeles malabarica, Aristolochia bracteolate, Calotropis gigantean, Clitoria ternetea, Gloriosa superba, Eclipta alba, Enicostemma axillare, Leucas aspera, Mimosa pudica and Sida cordifolia were still used by tribes and traditional healers as remedy for snake and poisonous bites (13,54-56).

Several recent studies have proved the weedy plants contain many medically useful active principles (alkaloids, glycosides, polyphenolics, steroids, tannins, resins, flavoniods, tetraploids and fatty acids) that are able to cure many nutritional disorders and diseases (1,57-67) in the human health care system.

CONCLUSION

This study points out that certain species of weeds are being exploited by the local inhabitants, but they are unaware the importance of all the plants in their agroecosystem. So, medicinally important species are easily discarded by the farming community. These weeds can become an additional source of income for the farmers, if they are made aware of the medicinal importance of these crop weeds. However there is a possibility of eroding this wealth of knowledge in the future is very fast due to less interest among the younger generation to protect ecosystems as well as their tendency to migrate to cities for lucrative jobs. Therefore, it becomes necessary to identification of specimens, proper documentation, awareness programs and introducing value addition activities related to processing of medicinally important plants through community enterprises are carefully designed to serve the needs of the community and introduced to younger generations in an effective way. This can be tremendous contribution to improving self reliance in primary health care for humans and gives supplementary income to the livelihoods and prevent the loss of our traditional plants and heritage.

The information regarding medicinal uses of plants reported are collected and scrutinized with published literatures and the farming community. However the therapeutic qualities and active principles of these plants should be standardized scientifically and tested for its safer use.

7

Indigenous Technical Practices in a Rice-based Farming Systems

INDIGENOUS CROPPING SYSTEMS

Indigenous cropping systems are those cropping systems that have been practiced for generations and still hold promise in meeting the food requirements of a growing population. Most of the cropping systems are well suited to the diversified agro-ecological conditions.

Sequential cropping is a system of cropping in which farmers sow two or three short duration crops in succession, especially legumes or oilseeds in lines between trees. Sequential cropping is adopted in marginal lands or dry lands. Sequential cropping contributes significantly to protein production for marginal and small-scale farmers.

Mixed cropping is a system of cropping in which farmers sow more than two crops at the same time. Farmers normally sow a mixture of legume and oilseed crops with an objective to meet protein and fat requirements. By sowing more than two crops, farmers try to avoid risks due to failure of any one crop. Mixed cropping is usually followed under rainfed conditions.

Monocropping is a system of cropping in which farmers cultivate the same crop in all three seasons in a year. Large-scale farmers who have access to irrigation prefer monocropping.

Intercropping is a system of cropping in which farmers cultivate two crops of different statures in alternate rows. Growing groundnuts and black gram or red gram and groundnuts are good examples.

Border cropping is a system of cropping in which farmers cultivate a major crop in the fields and a minor crop along the borders the fields. Rice and black gram are examples.

INDIGENOUS SOIL HEALTH CARE PRACTICES

Indigenous soil health care practices are those practices evolved, adopted, and modified by farmers based on their own informal experiments with an objective to maintaining the fertility and productivity of the soil. Crop rotation is a practice in which farmers grow different types of crops in various seasons. Crop rotation also implies that at least one legume crop should be incorporated in the cropping pattern in a year.

Fallowing is an indigenous soil health care practice in which farmers let cultivated land rest for a certain period of time before using it again.

Farm yard manure is a mixture of cow dung, cow urine, and paddy straw. Farmers apply farm yard manure especially to cereal crops such as rice, finger millets, and oilseeds such as groundnuts. Farm yard manure regulates the supply of nitrogen. Farm yard manure changes the color of the soil which is essential for absorbing sunlight. Farmers refer to this process as mann matram in Tamil.

Casuarina leaves: Farmers harvest Casuarina equisetifolia, a fuelwood tree, collecting the leaves and applying them to problem soils to counteract soil alkalinity.

Riverbed sand: Farmers apply sand that is collected from river beds if the problem of soil alkalinity is severe. There are some experienced farmers in the villages who can identify the severity of the soil alkalinity problem. Farmers facing alkalinity problems contact the experienced farmers for advise to correcting this problem.

Plowing Daincha in situ: Daincha is a root nodule shrub. Farmers with clayey soils, before planting rice, sow the Daincha seeds and plow the plants in situ when the plants become 45 days old. Mulching consists of leaving crop residues in the field, or bringing in other materials such as foliage from elsewhere.

Teprosia leaves: Farmers grow Teprosia populnea trees near the irrigation pump sheds. After second or third plowing, they cut the leaves of Teprosia and spread them over the plowed fields for one night. During the next day, they plow these leaves into the soil. After this operation, they puddle the field for planting.

INDIGENOUS RICE SEED SELECTION AND PROCESSING TECHNIQUES

Removing rogue plants: Rogues are different varieties of the same crop. Identifying and removing the rogue plants is a skillful technique. Farmers remove rogue plants at least 25 days before harvesting in order to avoid admixtures and also to maintain the genetic purity of a particular variety of a rice crop. Farmers claim that rogue plants mature first.

Spreading notchi leaves over the rice seeds: Once rice seeds are processed and stored, farmers spread notchi leaves over the rice seeds to prevent infestation by stored pests. Sieving rice seeds: Before sowing, farmers sieve rice seeds in order to separate the seeds of weeds. Since most of the weed seeds are bigger than rice seeds, they are filtered out in the sieves. Manual threshing of rice seeds: By threshing rice seeds manually, farmers claim that the plumule area of rice seeds are protected. Man farmers are of the opinion that tractor threshed rice seeds are of poor germination potential.

Selecting healthy plots: Farmers by physical observation demarcate a small plot for seed purposes. This is usually done one month prior to harvesting. Healthy plots that are free from pests or diseases attack are selected. Farmers also hold certain beliefs while selecting the rice seed plots. During the samba season, they select a plot from the north east corner of the field. This is locally termed as sani moolai. During the navarai season, they select a plot from the southwest corner of the field, locally termed as pillayar moolai.

Farmer-to-farmer seed exchange: Farmers practice their own system of obtaining quality seeds. They form an informal network wherein they visit each other's fields before harvest. They judge the quality of the seeds by observation. If they are satisfied, they buy from each other. There are some large-scale farmers in the village who raise one to two acre seed farms every season. Many small-scale and marginal farmers reported that these seed grower are more reliable than the public seed distribution system.

INDIGENOUS CROP NUTRIENT MANAGEMENT PRACTICES

Indigenous crop nutrient management practices are those manuring and fertilizing practices developed by farmers through judicious mixing of organic manures and chemical fertilizers.

Sheep manure: Some marginal farmers rear sheeps especially for their manure value. According to them, five to six sheeps are sufficient to cater to the manure needs of one acre of rice. Sheep manure is usually applied once in a year. Farmers who apply sheep manure usually skip the basal application of chemical fertilizers. Sheep manure is powdered and mixed with urea for top dressing. Sheep manure releases the nitrogen quickly when compared to farm yard manure. Farm yard manure: Farm yard manure is a mixture of straw, cow dung, urine, and other plant materials. Pure cow dung is not good for the rice crop. According to farmers, the farm yard manure has certain specific advantages: (1) Farm yard manure increases yield by at least two bags (1 bag=75 kgs.); (2) Farm yard manure increases the grain weight of rice; (3) Robust seedlings can be obtained by the application of farm yard manure; (4) Top dressing of nitrogen can be reduced if farm yard manure is applied basally; and (5) Farm yard manure adds roughness to the crop surface thus minimizing pest incidence.

INDIGENOUS RICE TRANSPLANTING TECHNIQUES

Row planting: Planting in rows significantly increases the production of tillers in rice variety ponni. This practice also enables the farmers to undertake intercultural operations such as application of fertilizers and pesticides. This practice yields an additional two bags of rice per acre.

Pinch planting: During navarai season, farmers ask the laborers to plant only 2-3 seedling per hill. This is locally referred to as killi poduthal (pinch planting). Clump planting: During samba season, farmers ask laborers to plant 45 seedlings per hill, This practice is locally referred to as pudichi poduthal (clump planting).

INDIGENOUS RICE WEED MANAGEMENT STRATEGIES

Indigenous rice weed management strategies are those strategies adopted by farmers to minimize the growth of weeds in rice fields.

Puddling rice nurseries followed by drying: Farmers irrigate the rice nurseries on the first day. This irrigation enables the weed seeds to germinate. They store the water for three-four days. Some farmers wait even for one week. The water slowly dries up leaving the weeds. Then they plough the fields by turning the soil upside down. Thus, the germinated weeds are killed. Again, they irrigate the nursery area and repeat the entire practice. Farmers claim that meticulous practice of alternative wetting and drying of rice nurseries helps them to minimize weeding.

Weed management in rice main field: The following management practices are being followed by the farmers in order to manage the weeds effectively in the main field: (1) Preparing and levelling the main fields uniformly without undulations; (2) Maintaining the heights of the field bunds at one inch; (3) Storing water continuously up to 15 days from planting. Draining of water especially during the first 20 days from planting leads to emergence of the weeds; (4) Maintaining the water level 1 inch; (5) Closer planting is necessary especially during the navarai and sornawari seasons; and (6) Applying neem cake to control korai weeds.

INDIGENOUS RICE PEST MANAGEMENT STRATEGIES

Pest monitoring: Most of the farmers apply pesticides after a thorough pest monitoring. Farmers look for pest symptoms in rice tillers. For each rice pest, farmers have their own economic injury levels. They apply pesticides only if the infestation crosses economic injury levels. Only for earhead bug, do they apply pesticide immediately even if only one bug is seen. Proper aeration to manage the attack of brown plant hopper: In order to minimize the attack of the brown plant hopper, farmers fold the rice crop once in eight feet. This practice not only provides aeration for the rice tillers but also exposes the culms of the rice crop where the brown plant hopper is usually found.

Local rat traps: Farmers invented these rat traps to kill rats in the rice fields. Rats are one of the major non-insect pests and contribute to 35% of grain loss. The damage is severe during the milky stage and grain formation stage. Farmers install these traps along the bunds to kill the rats during night times. The infestation is severe only during the night times. These traps are effective than chemical rodenticides. Moreover, rodenticides are known for polluting the environment as well as groundwater.

INDIGENOUS TECHNICAL PRACTICES FOR GROUNDNUTS

Sowing groundnuts in rice fallows: During the last week of December, after the harvest of samba paddy, utilizing the available moisture, farmers sow groundnut seeds using hand hoes.

Sowing of groundnuts normally completed within three to five days to utilize the moisture effectively. The following are the significant advantages of this practice over the conventional method of sowing the groundnut seeds after the field preparation:

1. Efficient use of time since land preparation for the next season is not required;
2. Saves cost of labor for the land preparation;
3. Infestation of Prodenia is below the economic threshold level;
4. Two weedings are sufficient; and
5. No significant difference in the yield.

Intercultural operation in groundnuts: Some farmers in Kizhur village turn the soil using hand hoes immediately after the second weeding of the groundnut crop.

This should be followed by pressing the soil closer to the groundnut plant up to a height of 3-4 inches from the surface of the soil. This practice results in increasing the rate of peg formation that consequently leads to higher yields in groundnuts. In spite of increased labor input for this activity, the farmers found a significant difference in the net profits.

DIFFERENTIAL SUPPRESSION OF RICE WEEDS BY ALLELOPATHIC PLANT AQUEOUS EXTRACTS

Weeds are one of the worst biological constraints to direct seeded rice cultures (Rao et al., 2007). Associated with the direct seeding of rice is an inevitable shift in weed flora to more difficult-to-control and non-native weeds in rice fields.

An appropriate weed control strategy comprised chiefly of herbicides has always been a major input and key element to sustain profitable rice production. However, the development of resistance in some previously susceptible weed species, as well as serious environmental concerns owing to high residual effects of herbicides in soil are major drawbacks associated with continuous herbicide usage (Ahn et al., 2005). Resistance of rice weeds to a number of herbicides has also been reported (Rao et al., 2007), probably because these target only a few molecular sites (Duke, 1990). In fact, the 270 herbicides covering the global market represent only 17 modes of action with almost half of them acting as ALS (acetolactate synthase), PS II (photosystem II), and Protox inhibitors.

Given the increasing emphasis placed on sustainable agriculture, and concern about the adverse effects of extensive use of farm chemicals, research attention is now being focused on reducing the dependence upon synthetic herbicides, and finding alternative strategies for weed management. Focused work on plant derived materials as an environmentally friendly alternative approach for weed control in field crops has been under way for the last two decades (Kuk et al., 2001).

Utilization of the allelopathic properties of native plant/crop species offers promising opportunities for this purpose. Many crop/plant species have been screened during the recent past for their allelopathic activity with an emphasis placed on weed management in agricultural systems. Allelopathic plants such as sorghum (Sorghum bicolor) (Cheema & Khaliq, 2000; Weston & Duke, 2003), sunflower Helianthus annuus), brassica (Brassica compestris), eucalyptus (Eucalyptus camaldulensis), mulberry (Morus alba) (Hong et al., 2004; Jabran et al., 2010), and winter cherry (Withania somnifera) (Jabran et al., 2010; Knox et al., 2010) are inhibitory to certain weeds.

Allelopathic potential of plant species can be exploited in many ways, and the utilization of the aqueous extracts of these species is one possible tool. Different plant species contain allelochemicals that vary in type and concentration (Xuan et al., 2005). Weeds can be better controlled by the utilization of plants that possess a greater fraction of allelochemicals (Elijarrat & Barcelo, 2001).

Identification of plant species with greater allelopathic potential, and the characterization of their adverse effects against weeds is required for better ecological based weed management. These plants can also prove to be a potential source of allelochemicals as a lead for natural herbicides (Vyvyan, 2002).

Little information is available regarding the phytotoxic effect, if any, of the plant extracts of sorghum, sunflower (brassica, eucalyptus, mulberry, and winter cherry against rice weeds. The suppressive allelopathic potential of these extracts was evaluated in pot studies on early seedling growth of rice weeds viz. horse purslane (Trianthema portulacastrum) [broad-leaf], jungle rice (Echinochloa colona) and E. crus-galli (barnyard grass) [grasses] and purple nut sedge (Cyperus rotundus) and rice flat sedge (C. iria) [sedges]. The agro-ecological significance of the results obtained for weed management in rice fields is also discussed.

MATERIALS AND METHODS

Preparation of extracts

Mature above ground foliage of sorghum, sunflower, brassica, winter cherry, leaves of mulberry, and eucalyptus plants were collected. These were stored and dried under shade to avoid possible leaching by rain water. To prepare the aqueous extracts, each plant material was chopped with an electric fodder cutter into 2-3 cm pieces. The chopped plant material was soaked in distilled water for 24 h at room temperature (25±5 oC) in a ratio of 1:10 (w/v) and was filtered through a 10 and 60 mesh sieve. Initial volume of the distilled water for soaking was 10 L that, after filtration, remained 7 L. The respective extracts were boiled at 100 oC to concentrate up to 20 times for easy handling and application. Previous studies revealed that boiling did not affect the nature, relative composition, or efficacy of allelochemicals (Jamil et al., 2009).

To exclude the notion of inhibitory effect due to the pH and osmotic potential of the plant aqueous extracts and skip possible concerns about the ecological implications of allelopathy, these two attributes were also measured for each of the aqueous extract. The pH and electrical conductivity of the extracts were recorded with digital pH and conductivity meters (HI-9811, Hannah, USA). The osmotic potential of different extract concentrations was computed as:

Osmotic potential (-Mpa) = Ec (ds m-1) x -0.036

Bioassay

Foliar application of allelopathic plant aqueous extracts was evaluated for their suppressive effects on the seedling growth of rice weeds. Plastic pots (29 x 18 cm, 10 kg capacity) were filled with air dried, sieved, well mixed soil (pH of saturated soil paste and electrical conductivity (EC) of the saturation extract were 7.9 and 0.41 d Sm-1, respectively). After germination, five seedlings (2-4 leaf stage, BBCH scale growth stage 12-14) of each weed species were maintained in each pot and placed in a screen house with natural solar radiation and an average temperature of 35±5 oC. Allelopathic plant aqueous extracts of sorghum, sunflower, brassica, mulberry, eucalyptus, and winter cherry each at 18 L ha-1 (60 mL L-1 of water) were sprayed on the 10th day after sowing with the help of a hand operated atomizer. Volume of the spray solution (300 L ha-1) was determined by using water. Control pots were sprayed with distilled water. These pots were irrigated as and when required to keep the soil moist and avoid water stress.

Both the root and shoot lengths were measured on the 14th day after foliar application of plant aqueous extracts. All roots and shoots from each pot were cut separately and oven dried at 70 oC for 48 h to obtain the dry biomass of root and shoot; total seedling biomass was calculated as the sum of biomass of root and shoot. Number of leaves, tillers, and/or lateral shoots/sprouts were counted manually and averaged. Chlorophyll content was measured after extraction in 80% ice cold acetone as per Lichtenthaler & Wellburn (1987), read by a UV-spectrophotometer (UV-4000, ORI, Germany), and are expressed as mg g-1 fresh leaf weight.

Experimental design and statistical analysis

Pots were placed in a factorial arrangement using a completely randomized design with four replications. Following an analysis of variance technique (Steel et al., 1997), the mean values were separated using the least significant difference (LSD) at P=0.05 employing MSTAT-C. Standard errors were calculated using MS-Excel.

RESULTS AND DISCUSSION

The results suggested the phyto-toxicity of allelopathic plant aqueous extracts against the seedling growth of weeds. A suppressive influence on the shoot and root lengths of weed seedlings varying in magnitude was enforced by almost all plant aqueous extracts used in this experiment. Significant (P=0.05) interactive effects for weed species and extracts were observed on all the seedling growth attributes depicting that allelopathic inhibition was species specific.

Eucalyptus inhibited the shoot and root lengths of horse purslane to a maximum of 36 and 20%, whereas mulberry and winter cherry promoted its root length by 24% over the control. Brassica and sorghum recorded a maximum (44 and 32%) reduction in shoot and root length of jungle rice, respectively. Mulberry aqueous extract recorded a maximum suppression in the shoot length of barnyard grass, while sorghum was most toxic to its root length. A perusal of the data also indicated that eucalyptus aqueous extract recorded the highest suppression in shoot length (60%) while brassica accounted for maximum inhibition in root length (56%) of purple nut sedge. Shoot length of flat sedge appeared more susceptible to sorghum (53%) while its root length was more inhibited by eucalyptus (64%) aqueous extract.

The seedling shoot and root lengths have been the most frequently used parameters for the expression of allelopathic potential (Inderjit & Dakhshini, 1995; Prately et al., 1999). The inhibition of shoot and root growth in the tested weed species upon exposure to allelopathic plant aqueous extracts may be attributed to phytochemicals present in these extracts. The pH and osmotic potential of different extracts ranged between 6.20 to 6.80 and -0.50 bars to -0.88 bars, respectively, and both were unlikely to avert plant growth, and growth inhibition was thought presumably due to inhibitory compounds present in the aqueous extracts. Netzly & Butler (1986) and Cheema et al. (2009)

reported several phytotoxins such as gallic acid, protocatechuic acid, syringic acid, vanillic acid, p-hydroxybenzoic acid, p-coumaric acid, benzoic acid, ferulic acid, m coumaric acid, caffeic acid, p-hydroxybenzaldehyde and sorgoleone in sorghum.

Sunflower also contains allelochemicals viz. chlorogenic acid, isochlorogenic acid, a naphthol, scopolin, and annuionones. The members of the Brassicaceae family contain glucosinolates that, upon hydrolysis, yield isothiocyanates, and isoprenoids and benzoids that exert allelopathic effects on the germination and growth of other species. Eucalyptus species are considered to be one of the most notorious allelopathic trees and release sufficient quantities of terpenes, phenylpropanoids, quinones, coumarins, flavonoids, tannins, phenolic acids, glycosides, and cyanogens (Einhellig, 1986). Allelochemicals of mulberry and winter cherry have remained undiscovered until now although their allelopathic potential has also been reported in some recent studies by Jabran et al. (2010).

Reduced root and shoot lengths of test weed species can be due to alterations in DNA synthesis in their respective apical meristems, mitochondrial metabolism or changes in cell mitotic indices or a combination of all of them. Al-Wakeel et al. (2007) reported that retarded seedling elongation might be an outcome of the direct interference of allelochemicals with the process of cell division that alters the balance of different growth hormones. Differences in the activity of different extracts in suppressing the seedling growth of the tested weed species may be due to difference/s in the type and concentration of allelochemicals present in these extracts (Xuan et al., 2005).

A reduction in leaf score and lateral branches/tillers/side shoots of tested weed species indicated that the size of photosynthesizing machinery was also drastically suppressed. Eucalyptus and mulberry accounted for the greatest suppression of these characteristics in horse purslane (51 and 47%), and jungle rice (49 and 46%). Winter cherry was the most suppressive against leaf score (66%) and mulberry against tillering (63%) in barnyard grass. Sunflower and sorghum appeared more inhibitory to purple nut sedge (45 and 67%) and flat sedge (50 and 65%), respectively.

Likewise, there was a reduction of fresh and dry biomass of weed seedlings relative to control under the influence of all the aqueous extracts. However, again a species specific response regarding both extract sources and tested weed species was indicated by the presence of a significant interaction (P=0.05) between these. Eucalyptus aqueous extract was more effective in retarding the dry matter accumulation of horse purslane (49%) and jungle rice (59%) seedlings while winter cherry scored maximum (>60%) fresh and dry weight suppression of barnyard grass. Sunflower was the most detrimental to purple nut sedge and sorghum and brassica aqueous extracts to flat sedge.

The total chlorophyll content of weed leaves treated with allelopathic plant aqueous extracts declined compared to the control and variable levels of reductions were observed for different weeds species under the influence of different extracts.

Disturbance and alteration in seedling morphology under allelopathic stress in our studies can be explained by alteration in mitochondrial respiration thus yielding less ATP for energy consuming processes (Gniazdowska & Bogatek, 2005). Allelochemicals in plant aqueous extracts might have inhibited chlorophyll in the tested weed species by interfering with the biosynthesis of photosynthetic pigments or enhancing their degradation or through the integration of both (Huang et al., 2010). Changes in chlorophyll contents in our studies are also supported by the findings of Inderjit & Dakshini (1992) who cited allelochemical-mediated reduction in seedling photosynthetic pigments primarily due to phenolic acids.

Although the exploitation of allelopathy as a natural weed management tool is eye-catching as an eco-friendly weed management approach yet contrary to bioassays, the suppression achieved under field conditions is often negligible (Duke et al., 2001). Many studies have obtained contrasting results under controlled and natural conditions (Inderjit & Weston, 2000). Nonetheless, soil possesses the ability to detoxify allelochemicals, so the bioassays conducted under controlled conditions in the absence of soil might be misleading due to an overestimation of the allelopathic potential (Foy, 1999; Inderjit, 2001). It is imperative to conclude whether these allelochemicals can accumulate under field conditions and affect individuals of a weed community.

This confirms the need to carry out field trials to quantify suppression caused by an allelopathic species. It is often impossible to simulate the exact field conditions in laboratory trials, but an effort was made in the present studies by using soil as a growing medium with the objectives of maintaining the physical, chemical, and biological soil factors of the natural setting.

Inderjit (2001) also discussed the significance of soils in the expression of allelopathy that may be explained under field or pot experiments. Foy (1999) concluded that in terrestrial systems any bioassay without involving soil has no ecological relevance. Moreover, the pots were placed under screen house and under open environments.

Despite their known limitations and criticism received, bioassays are useful tools as early proof for allelopathic potential. Our data showed strong inhibitory potential of different species against the seedling growth of tested weed species. Seedling biomass can be used as a bio-indicator for the inhibitory effect of allelopathic extracts. Although the reduction in seedling biomass is specific for both donor and receiver species, but the aqueous extracts of sunflower and eucalyptus were more suppressive to a number of weeds, and may be evaluated by applying either at a higher concentration or in combination in order to cope with the diversity of weed flora in rice fields. Efforts are needed to identify and isolate the most effective allelochemicals from these species in a quest for natural herbicide molecules.

SEED WINTERING AND DETERIORATION CHARACTERISTICS BETWEEN WEEDY AND CULTIVATED RICE

Traditional rice cultivation has been via transplanting of rice seedlings. However, due to the severe problems of agricultural water scarcity and farm labor shortage, direct-seeding methods have been practiced in many parts of the world. In most countries, weedy rice infestation increased significantly after shifting from rice transplanting to direct seeding, and it was recognized as a noxious weed. Weedy rice infestations have been reported to have spread to 40–75% of the total area of rice cultivation in Europe, 40% in Brazil, 55% in Senegal, 80% in Cuba, and 60% in Costa Rica.

The occurrence of weedy rice has very large effects on rice yield. Weedy rice infestations are responsible for significant yield losses, which are particularly severe in short varieties and late planting cultivation. Weedy rice is so productive that it can spread and cause major economic damage, but it is impossible to control weedy rice during the rice cultivation period by herbicide because it belongs to the same species as the cultivated rice.

Weedy rice looks similar to cultivated rice in outward appearance, but has a high level of adapting ability to environmental stress in physiological traits. Being distinguishable by a red pericarp, it disperses immediately by shattering at maturity, guaranteeing its continuation. The shattering nature of weedy rice after ripening makes it difficult to be removed when cultivated rice is harvested. The occurrence of weedy rice becomes more severe the following year due to shattered seed present in the soil.

The reason of continuous incidences of weedy rice was that its shattering seeds were able to over-winter. In this respect, the deterioration of weedy rice seeds during winter might be different with cultivated rice in freezing resistance and antioxidant activities. There were no reports on the freezing resistance, but some reports of antioxidant activities related with seed deterioration in rice seeds. Generally, it has become increasingly accepted that damage resulting from reactive oxygen species (ROS) or oxidative stress plays a role in the seed aging process. ROS are highly reactive and may modify and inactivate proteins, lipids, DNA, and RNA and induce cellular dysfunctions.

Plants require high contents of antioxidants such as ascorbate, a-tocopherol, glutathione, and phenolic compounds that can remove ROS as a defense mechanism, and enhance the activities of antioxidant enzymes like superoxide dismutase (SOD), peroxidase (POD), catalase (CAT), and ascorbate peroxidase (APX). If balances are broken between ROS generation and removal in plant cells, oxidant losses occur along with damage to lipid peroxidation and cell membranes.

In the present study, the wintering of shattered seeds of weedy rice on paddy fields was compared with cultivated rice, and seed deterioration and related characteristics concerning seed wintering are reported.

RESULTS

Seed viability of weedy rice (WD-3) and cultivated rice (Hopum) after a wintering test on the surface of a paddy field. The viability of wintered seeds was 92.7% for the weedy rice and just 4.3% for the cultivated rice. Wintering characteristics of weedy rice in the winter period may vary depending on the moisture state of the paddy surface. Therefore, the second field experiment for seed wintering was carried out in a flooded paddy as well as in a dry paddy field. The results were almost identical to the first year experiment except that the seed viabilities significantly decreased in the flooded paddy as compared with the dry paddy field for both weedy rice (WD-3, PBR) and cultivated rice (Hopum, Ilpum). The seed viability of cultivated rice rapidly decreased from December to February regardless of paddy conditions and the final average viabilities after wintering were 5.0% (4.3% in Hopum, 5.7% in Ilpum) in the dry paddy and 0.5% (0.3% in Hopum, 0.7% in Ilpum) in the flooded paddy. For weedy rice, the viability was maintained over 90% in the dry paddy, but decreased slightly during winter to a mean of 61% (WD-3: 76%, PBR: 58%) in the flooded paddy field. At the early stage of wintering, the germination percentage of weedy rice was much lower overall than of cultivated rice.

This may be because weedy rice had primary seed dormancy, and generally exhibited stronger dormancy than cultivated rice. In considering the relationship of seed dormancy to wintering, a second area of interest is whether dormancy increases the lifespan of seeds. In this result, the dormancy of all seeds disappeared in December regardless of its intensity and did not affect the resistant to wintering and freezing stress. Roberts reviewed on the subject and concluded that available evidence was not sufficient to establish even a causal relationship. However, this does not exclude the possibility that the lifespan of dormant seeds of native or wild plants buried in the soil under natural conditions may be longer than the lifespan of non-dormant seeds.

Viability of seed wintered on the paddy surface may be affected by temperature changes. In particular, subzero temperatures after rain and snow may freeze the seeds and cause them to die. Therefore, in order to winter safely under subzero temperature, seeds should have resistance to freezing injury for not losing its viability even under soaking conditions. The seed viability after a number of freezing treatments (1–3?cycles). Seed viability values on average of three treatments of freezing were about 78% (PBR: 72%, WD-3: 85%) in weedy rice and 16% (Hopum: 3%, Ilpum: 29%) in cultivated rice. The results indicated that the weedy rice had much higher resistance than the cultivated rice varieties, and there were also considerable variations among accessions regardless of the number of freezing treatments.

The deterioration of seed tissue induced by freezing treatment was observed by the tetrazolium test. The frequency of dark purple stained seeds matched up with the freezing resistance of 4 accessions. In most seeds of weedy rice (PBR and WD-3), whole embryo tissue and the aleuronic layer were stained dark purple.

In the cultivated rice, however, the seeds of Hopum were not completely stained, but those of Ilpum were partially stained, which results indicated that freezing resistance of Ilpum was higher than that of Hopum, and there were variation of freezing resistance among individual seeds in a variety and among rice varieties. Some seeds partially damaged were stained in a light pink at overall tissue such as radicle, plumule, scutellum in embryo and aleurone layer in endosperm. In the deterioration process of seeds under subzero temperature, there was no special tissue structure initiating deterioration.

During the winter season, shattered seeds on the paddy surface tended to remain wet at low temperature. Therefore, an accelerated aging test was set up by soaking the seeds at a low temperature (4°C). In this condition, the change of seed viability, physiochemical properties including protein content, fat acidity, and some antioxidant enzyme activities were observed at 10?day intervals during treatment. The seed viability of weedy rice under the accelerated aging condition started to fall off 50?days after treatment and reached approximately 40% 90?days after treatment. In the cultivated rice, it started to fall off 20?days after treatment, and reached 4% 90?days after treatment.

The time courses of protein content and fat acidity in seeds during accelerated aging treatment. The protein contents of WD-3 and Hopum, which were 7.41% and 6.44% before treatment, respectively, significantly decreased to 5.25% and 4.82% 40?days after treatment, and then maintained at this amount until 90?days after treatment without significant change.

In the case of fat acidity, the initial values were 6.50 KOH mg/100?g in the weedy rice and 6.59 KOH mg/100?g in the cultivated rice, which were the same between the two varieties. Until 40?days after treatment, there were no significant changes of fat acidity, but there was a significant increase and difference between weedy rice and cultivated rice 50?days after treatment. Fat acidities of weedy rice and cultivated rice were 42.9 KOH mg/100?g and 23.9 KOH mg/100?g at day 70, and 44.1 KOH mg/100?g and 30.0 KOH mg/100?g at day 90, respectively.

The change of superoxide dismutase (SOD) activity during accelerated aging treatment. The initial values of SOD in the seeds were 42.9 units/mg protein in WD-3 and 11.7 units/mg protein in Hopum. Following accelerated aging treatment, SOD activities of seeds increased for 10?days to 104.9 units/mg protein in WD-3 and to 19.9 units/mg protein in Hopum. After peak SOD activity, they decreased in both weedy and cultivated rice until the end of treatment. The change of SOD activity of weedy rice seed rapidly increased and decreased compared to Hopum, but it was maintained at higher levels than that of Hopum. At day 90 of accelerated aging treatment, the SOD activities were 14.3 units/mg protein in weedy rice and 6.4 units/mg protein in cultivated rice. The results of SOD activity changes during accelerated aging suggested that the response of weedy rice seed to environmental stress was more sensitive than cultivated rice.

Change in catalase (CAT) activity in the seeds. The initial CAT activity of WD-3 was much higher than that of Hopum. They were 26.9 units/mg protein in weedy rice and 6.6 units/mg protein in cultivated rice. After accelerated aging treatment, CAT activities gradually decreased in both weedy and cultivated rice to 3.9 units/mg protein in weedy rice and 0.8 units/mg protein in cultivated rice at day 90 of accelerated aging treatment. The CAT activity of weedy rice seed persisted at a significantly higher level than that of cultivated rice seed during accelerated aging treatment.

Free radical scavenging of phytochemical substances in seeds is an important and generally accepted mechanism for antioxidants inhibiting oxidation. The results of seed antioxidant activity measurement during accelerated aging using DPPH. The initial radical elimination activity of WD-3 seed was 93.0%, which was much higher than the 56.6% of Hopum. This persisted during accelerated aging for 90?days without major changes, with approximately 90% in weedy rice and 50% in cultivated rice.

DISCUSSION

In this study, the wintering rate of weedy rice seeds was about 90% on the dry paddy and 60–80% on the flooded paddy, which was much higher than that of cultivated rice (below 5%). This result was in agreement with several reports about the high germination percentage of weedy rice seeds placed in shallow depths or on the soil surface of paddy fields in the winter. Therefore, if a seed of weedy rice spread, it could occur every year and exponentially increase in the rice field. Fogliatto et al. highlighted the significant efficacy of winter flooding in reducing weedy rice infestations in paddy fields. This report is meaningful because the practice showed a significant reduction of weedy rice seed density compared to overwintering under dry conditions, and also winter flooding could favor the decay of seeds of several weeds including weedy rice. However, results reported here indicated more than 60% of weedy rice seeds could overwinter under flood conditions. Therefore, winter flooding may not be completely effective but may be an accompanying method to control weedy rice in Korea.

In the winter season, the major factors affecting seed longevity in the field may include precipitation and subzero temperature, as shown in meteorological data. When the plant is being frozen gradually, moisture leaves the cell to the outside of the membrane and forms ice crystals causing mechanical damage. In the case of seed, if its moisture content is high, it is easily deteriorated in above freezing temperature owing to an increase in respiration and metabolism, and may get damaged in the winter or die at subzero temperature.

The freezing resistance of weedy rice seed was much higher than that of cultivated rice. Weedy rice surviving semi-wildly in rice fields is known to show a high level of adaptability to environmental stress. The weedy seeds that shatter on the surface of paddy fields in the autumn should have resistance to freezing in the winter to regenerate in the following germinating season, although there was no report on the freezing resistance of weedy rice seeds. In plants with freezing resistance, antifreeze proteins block the growth of ice crystals in the outside spaces of cells in the tissue, thus preventing cell damage derived from the repetition of freezing and thawing, and also have functions in transforming the generation of ice crystals by adhering to surfaces of ice. In this way, antifreeze protein accumulates in the tissue in case of cold acclimation, and thus makes the plant not freeze to death even in freezing temperature. The tetrazolium test, that freezing damage of the cultivated rice began in overall tissue of the embryo and aleuronic layer of seed. But, in the weedy rice seeds after freezing treatment, those seed tissues were stained by tetrazolium solution to dark purple to show freezing resistance. Therefore, one hypothesis for a freeze resistance mechanism of weedy rice is that anti-freeze proteins accumulate in the embryo and aleuronic layer of the seed.

The accelerated aging test has been recognized as an accurate indicator of seed vigor and longevity, and may be generally induced by high temperature and high humidity. The deterioration of seeds wintering on the field surface in temperate regions could be affected by high moisture content at low temperature as discussed earlier, thus accelerated aging can be induced by soaking at low temperature (4°C). The seeds deteriorated rapidly under this accelerated aging method, and there was a large difference in deterioration rate between weedy and cultivated rice. The viabilities of seeds after 90?days of accelerated aging were 4% in cultivated rice and 40% in weedy rice.

The deleterious effect of accelerated aging on the longevity of seeds was associated with the damage occurring to lipid, nucleic acid, and protein owing to oxidative stresses. In the accelerated aging test, the difference of protein contents and fat acidities in seeds between weedy and cultivated rice indicated that weedy rice had a higher protective system to aging stress than the cultivated rice.

Scandalios reported that SOD and CAT were effective antioxidant enzymes for decreasing oxidant damage of cells by oxidative stress. The activity of SOD before and after the accelerated aging treatment for 90?days in weedy rice was much higher than in cultivated rice. However, the SOD activities of seeds rapidly increased and decreased with the peak occurring 10?days after treatment.

SOD activity decreased if stresses persisted for a long time or excessive stresses were added. CAT is considered a primary enzymatic defense against oxidative stress induced by senescence, chilling, dehydration, osmotic stress, wounding, paraquat, ozone and heavy metals. Shon et al. reported that CAT activity had exhibited a decreasing tendency upon submerging stress in young rice seedlings, and appeared as the most sensitive enzyme to stress among antioxidant components, regardless of whether it was in seeds. Similarly, CAT activity was reduced after treatment in both the weedy and cultivated rice, where the activity of weedy rice persisted at a higher rate than the cultivated rice. Tanida reported that CAT activity of the embryo in the seed soaked at low temperature had positive correlation with germination percentage after treatment. High SOD and CAT activities, as well as radial scavenging activity were considered to increase the vitality of seeds from excessive stress damage in soaking at low temperature.

METHODS

PLANT MATERIALS

Two japonica weedy rice, PBR and WD-3, and two Korean bred cultivars (japonica), Hopum and Ilpum, were used in this experiment. Prior to the laboratory-based experiment, they were regenerated in the experimental field in Chonbuk National University to obtain fresh seeds. The weedy rice were harvested at 30?days after heading, and the cultivated rice were harvested at 50?days after heading. Seeds were air dried until seed moisture content (SMC) became less than 14%. The SMC was measured by oven-drying method. The seeds were sealed in plastic bags, and stored at 4°C until use for about 7–8?months. When they were used, the SMCs were about 12% regardless of varieties.

SEED WINTERING TEST ON THE PADDY FIELD

In the first wintering experiment, Hopum, a bred cultivar, was cultivated in the dry paddy field where the weedy rice (WD-3) occurred at a rate of more than 500 plants per m2. After harvesting Hopum, naturally shattered weedy rice (WD-3) seeds and artificially shattered cultivated rice (Hopum) seeds were placed on the soil surface at 3 spots of the paddy field under wire net protection during winter from November 2008 to April 2009. Germination tests of seeds collected at 3 spots after wintering were conducted at 25°C for 14?days using 4 replicates of 100 seeds each in accordance with the International Rules for Seed Testing for Oryza sativa.

In the second wintering experiment, two weedy rice varieties (WD-3 and PBR) and two Korean cultivars (Hopum and Ilpum) were tested. The seeds of 4 varieties (200?g each) were kept on the paddy from November 2009 to April 2010. The paddy conditions were maintained as two types during winter. One was a dry paddy as was used for the first wintering test and the other was a flooded paddy simulated by plastic boxes (41.0?×?24.5?×?15.0?cm) filled with 10?cm depth of paddy soil (silty clay loam) and superabundant water. During wintering, the seeds (400 grains each from three replicates) were sampled 5 times for viability tests. The meteorological data on temperature, subzero temperature days, precipitation, and precipitation days during the wintering test for 2008/2009 and 2009/2010.

SEED FREEZING TEST

To investigate the resistance of seeds under subzero temperature, the weedy (PBR, WD-3) and cultivated rice (Hopum, Ilpum) were assigned to freezing treatment as follows; the seeds were frozen at -10°C for 24?hours after imbibing at 4°C for 24?hours and then thawed at 4°C for 24?hours. The freezing method was repeated 1 to 3 times, and then the viability of the seeds was tested as stated above in seed wintering test. After freezing treatment, a tetrazolium test was also carried out to investigate the state of seed tissues. Soon after freezing treatment, seeds were kept at 25°C for 15–18?hours. Following that, seeds and embryos were cut vertically using a razor blade, placed in 1% tetrazolium solution (pH 6.5–7.5) at 30°C for 2?hours, and the staining patterns were observed under a dissecting microscope.

SEED VIABILITY AND PHYSIOCHEMICAL TEST UNDER ACCELERATED AGING TREATMENT

To determine the deterioration of soaked seed at low temperature, the seeds of weedy rice (WD-3) and cultivated rice (Hopum) were kept soaking in distilled water at 4°C for 90?days as an accelerated aging treatment, and viability and physicochemical changes of seeds were observed at 10?day intervals. During accelerated aging treatment with three replicates, a 100?g of seeds per replicate was sampled at each sampling time. Among the sampled seeds, 400 grains per replicate was used for the viability test and the rest was ground with liquid nitrogen to powder by auto-mill (TK-AM5, Tokken Inc.)

and then dried by freeze dryer (Clean Vac 8, BioTron). The powder was used to analyze protein content, fat acidity, superoxide dismutase (SOD) activity, catalase (CAT) activity and 2,2-diphenyl-2-picrylhydrazyl (DPPH) radical scavenging activity. A germination test was conducted as for the wintering test with four replicates.

Protein content was measured in three replicates by the micro-Kjeldahl method. Seed powder (0.2?g per replicate) was weighed into the Kjeldahl flask. Sulfuric acid and a catalyst (Kjeltabs auto: 1.5?g K2SO4, 7.5?mg Se) were added to the flask and the mixture was heated to 420?°C for 1?hour. Total nitrogen of each protein fraction was determined by the automatic Kjeldahl nitrogen analyzer and protein content was calculated using a conversion factor of 5.95, as shown in the equation below:

Protein(%)=N(%)×5.95(1)

Fat acidity (KOH mg/100?g) was analyzed in three replicates by the AOAC method. It was determined by the amount of KOH required to reduce free fat acid included in 100?g of dry material. A total of 10?g of seed powder was placed in a 100?ml flask followed by addition of 50?ml of benzene and 30?min. agitation (150?rpm, 25°C). The mixed solution was filtered through filter paper, and 15?ml of the solution was put in a 100?ml flask and treated with 15?ml phenolphthalein melted in ethyl alcohol. The solution was finally titrated with 0.0045?N-KOH. Ending point of the reaction was determined when color of the solution turned pink, which was the standard color. Tests of the blank and standard varieties were carried out at the same time. Fat acidity was calculated by the following formula:

Fatacidityvalue=(T-B)×8.33×100/(100-W)(2)

T: amount of KOH (ml) used in sample

B: amount of KOH (ml) used in blank

W: moisture content of sample

8.33: conversion factor according to sample amount and KOH concentration

For analysis of SOD and CAT activity, 1?g of seed powder was homogenized with 5?ml 0.1?M potassium phosphate buffer (pH 7.0) contains 1?mM EDTA and 1?mM DMSO. The homogenized samples were centrifuged at 30,000?×?g for 20?min. at 4°C. The supernatant was stored at -80°C and used as a crude enzyme extract. For SOD activity test by the nitro blue tetrazolium (NBT) reduction method, 0.1?ml of 400?mM methionine, 0.1?ml of 2.5?mM NBT and 0.3?ml of 1?mM EDTA were vortexed in the glass test tube at 25°C for 3?min. for temperature balance. Then, 0, 20, 40, and 60?μl of enzyme extracts and 2.45, 2.43, 2.41 and 2.39?ml of 50?mM potassium phosphate buffer (pH 7.8) were added respectively to make final volume of 3?ml of enzyme and potassium phosphate buffer. 50?μl of 120?μM of riboflavin was added to each test tube. The test tubes were softly shaken and placed under fluorescent lamps for 15?min., and absorbance was measured at 560?nm by a spectrophotometer. One unit of SOD was defined as the amount of enzyme necessary to inhibit formation of blue formazan to reduce absorbance to 50% of blank's.

CAT activity was measured in accordance with Beers and Sizer. One ml of 59?mM H2O2(Merck's Supersol) and 1.9?ml of 50?mM potassium phosphate (pH 7.0) were put in the quartz cuvette. 0.1?ml of enzyme extract was added. The absorbance was measured at 240?nm using a spectrophotometer adjusted to 25°C. Finally, decrease in absorbance for 3?min was recorded. One unit decomposes one micromole of H2O2 per minute at 25°C and pH 7.0. The catalase activity was obtained by the following equation:

Units/mgprotein = (?A240/min × 1000)/(43.6 × mgenzyme/mlreactionmixture)

DPPH free radical scavenging activity was measured according to Lee et al. based on Blois method. A total of 1?g of milled seed sample was extracted by 10?ml methanol at 50°C for 24?hours on a 150?rpm shaker. The filtered sample solution was centrifuged at 12,000× gfor 20?min, and the supernatant was used for the assay. The reaction mixture containing 1,700?μl of 0.15?mM DPPH in methanol and 300?μl of sample solution was shaken well and incubated for 30?min. at room temperature, and the absorbance of the resulting solution was read at 517?nm. The scavenging activity was determined by comparing the absorbance with that of the control containing equal volumes of DPPH solution and ethanol. The DPPH radical scavenging activity was obtained by the following equation:

DPPHradicalscavengingactivity(%) = [(Acontrol-Asample)/Acontrol] × 100)(4)

STATISTICAL ANALYSIS

For the evaluation of individual treatment means, all collected data from the complete randomized design with three replicates (four replicates in germination test, exceptively), subjected to analysis of variance using Statistical Analysis System (SAS 9.1).

8

Sequence Polymorphisms in Wild, Weedy, and Cultivated Rice

INTRODUCTION

Domestication is the anthropogenic process by which a wild species evolves into new forms that meet human needs. Plant domestication has created hundreds of crops that have contributed tremendously in human civilization (Diamond 2002). A handful of agro-morphological and physiological characteristics, such as reduced seed shattering and dormancy, increased grain size, and synchronization of flowering and seed maturation, are essential for the evolution of domestication of major grain crops (Sang and Ge 2007). Consequently, this suite of characters is commonly referred to as the "domestication syndrome" (Harlan 1975). Perhaps, the most important of these characters is reduced seed shattering determined by genetic changes in the development in the abscission layers at the base of seeds or spikelets (Fuller et al. 2009). Therefore, studies elucidating the genetic basis of reduced seed shattering are critical for understanding the evolutionary process of cereal crop domestication.

Asian cultivated rice is the world's most important cereal crop, providing the staple food for more than one half of the global population (Khush 2005). It is also one of the earliest domesticated crops. Archeological evidence suggests that domestication of O. sativa from wild O. rufipogon Griff. occurred in the middle and lower Yangzi River regions of China about 8000 years ago.

As for other grains, a key evolutionary change during rice domestication involves the reduction in seed shattering. The infructescences of both Asian rice's wild ancestor (O. rufipogon complex, including the annual O. nivara) and Asian rice's weedy descendant ("weedy rice" also referred to as "red rice", Oryza sativa f. spontanea) shatter easily to disperse their seeds. Human selection for seeds that not dispersed, but retained on the plant, was essential for increasing grain yields of hand-harvested rice.

Fig. Panicles of cultivated rice (Oryza sativa, back) and weedy rice (O. sativa f. spontanea, front) that co-occurs with in agricultural ecosystems. Owning to its strong seed shattering at maturity, weedy rice causes a considerable yield loss of the crop rice.

Fig. Spikelets of perennial common wild rice (Oryza rufipogon). This grass species with strong seed shattering is considered the direct ancestor of Asian cultivated rice domesticated ~8000 years ago.

The genetic basis of seed shattering in rice has been examined in studies that identified the co-segregation of the trait and molecular genetic markers in progenies descended from crosses between cultivated rice and its wild relatives. Such studies have identified several quantitative trait loci (QTL) that play a role in whether rice seeds shatter from or persist on infructescences, such as sh3, sh4, sh6, sh8, qsh1, qsh2, qsh5, qsh11,qsh12, and sh-h (Konishi et al. 2006; Li et al. 2006; Ji et al. 2010).

Two shattering QTL, sh4 located on chromosome 4 and qsh1 located on chromosome 1, often considered to be critical for rice domestication and evolution, have been cloned. The sh4 locus that explains as much as 69% of phenotypic variance in shattering (Li et al.2006) was identified based on crosses between cultivated rice and wild relatives, including its ancestors in theO. rufipogon complex. Sequence data showed a single nonsynonymous substitution [a nucleotide substitution of "G" to "T" or an amino acid substitution of asparagine for lysine (Li et al. 2006)] in the first exon of sh4 that co-varied with the nonshattering phenotype. Transformation experiments by Li et al. (2006) demonstrated that this substitution "undermined the gene function necessary for the normal of the abscission layer that controls the separation of a grain from the pedicel", converting rice plants from shattering to a considerable degree of seed persistence (Li et al. 2006). This "G" to "T" substitution has been found to be fixed in all rice cultivars examined to date (Li et al. 2006; Lin et al. 2007; Zhang et al. 2009; Thurber et al.2010). Therefore, sh4 is widely considered to be the most significant shattering gene that is responsible for the domestication of rice from its wild ancestor (Li et al. 2006; Purugganan and Fuller 2009).

The other locus, qsh1, was identified in an F2 population derived from a cross between an indica rice cultivar and a japonica rice cultivar and explained almost 69% of phenotypic variance in shattering in that population. Again, sequencing data also showed a single nonsynonymous substitution ("G" to "T"), in this case, in the 5' upstream regulatory region of the locus that further reduced shattering ability in a subset of japonica rice cultivars. However, the evolutionary interpretation of QTL depends on the relationship of the parents in the cross that generates the segregating population. In the case of qsh1, the polymorphism only involves seed-shattering variation within cultivated rice. As acknowledged by those who cloned it, this allele could not play a role in rice evolution until after the early stages of domestication. Indeed, the role ofqsh1 in rice evolution is now considered to be important during the improvement of japonica varieties, rather than evolution during the first steps of the evolution of domestication rice.

The current view then is that the initial origin of nonshattering during the early rice domestication involved a single gene of large effect, specifically, the sh4 locus, with the evolutionary replacement of a dominant allele for shattering with a recessive allele for nonshattering. To date, most published sequence data of the sh4locus are consistent with that hypothesis involving the replacement of the wild type, with the seed-shattering "G" nucleotide, by the mutational type, with nonshattering "T" nucleotide, at sh4's functional nucleotide polymorphism (FNP) site as the key event in the domestication process of Asian cultivated rice (Li et al.2006; Lin et al. 2007; Zhang et al. 2009; Thurber et al. 2010).

However, some data challenge that hypothesis. A few studies have reported that some accessions of Asian rice's wild ancestor (O. rufipogon) contain the sh4 "T" substitution despite their seed-shattering phenotype. Furthermore, highly shattering weedy rice in the United States, a descendant of cultivated rice, appears to be fixed for the cultivars' "T" type. Finally, the phenotypic strength of cultivated rice's sh4's genotype varies almost fivefold (from as little as 15% to as much as 69% of phenotypic variation) depending on the parents used for QTL analysis (Cai and Morishima 2000; Li et al. 2006).

Collectively, these data suggest that this mutant allele at the sh4 locus did not necessarily play a significant role in the evolution of reduced shattering in the early domestication of cultivated rice. Clearly, the sh4polymorphism plays some role in shattering, but it is not at all clear that it played the major role for the domestication of Asian cultivated rice. The sh4 locus hypothesis depends on the rare mutant "T" allele replacing wild "G" allele as the earliest step in the domestication of rice. If the sh4 locus hypothesis is correct, then given the extreme rarity of nonshattering phenotype in wild populations, the recessive allele should be «1% in frequency in wild populations and, when homozygous, should express a nonshattering phenotype.

But we posit instead that one or more other loci contributed significantly to preventing shattering in wild rice in its early transition to domesticated rice. If our hypothesis is correct, then we would expect that populations of rice's wild ancestor should be polymorphic for sh4's genotypes, that the "T" allele should not be extremely rare, because other loci also play significant roles in the expression of shattering and nonshattering. However, prior studies have not surveyed the O. rufipogon complex sufficiently to characterize the sh4 polymorphism. Thus, further investigation is necessary involving sh4 screening of a sufficient number of wild rice samples of diverse geographic origins.

We tested the hypothesis of the sh4 locus as key to the domestication of cultivated rice by analyzing the sh4sequences of a large number of diverse wild, weedy, and cultivated rice samples. We identified those wild and weedy rice samples with strong seed shattering that had the putative "nonshattering" sh4 genotype ("T" nucleotide at the FNP site). Our primary objectives were to answer the following questions: (1) Is the sh4"wild-type" allele (i.e., with the "G" nucleotide at the FNP site) fixed or nearly fixed in wild and weedy rice accessions? (2) If sh4 is polymorphic in wild and weedy rice accessions, how do the "G" genotype and "T" genotype covary with the seed-shattering phenotypes of wild, weedy, and cultivated rice? Answering these two questions will allow us to address the larger question of whether the sh4 shattering locus has played a significant role in the domestication of Asian cultivated rice.

MATERIALS AND METHODS

SAMPLE COLLECTION AND PHENOTYPIC CHARACTERIZATION

Our germplasm panel consisted of a total of 76 wild rice accessions, 165 weedy rice accessions, and 125 cultivated rice accessions widely collected from Asia, southern Europe, and North America. We randomly chose one individual to represent each accession. In addition, we supplemented our data with the published DNA sequences of sh4 from 90 wild rice accessions, 57 weedy rice accessions, and 67 cultivated rice accessions (Li et al. 2006; Zhang et al. 2009; Thurber et al. 2010).

The phenotypes of seed shattering versus persistence were determined by gently gripping each of three naturally mature panicles per plant by hand. If more than 60% seeds were released from each panicle by this treatment, a plant was determined to have a "seed-shattering" phenotype; if less, the phenotype was characterized as "seed persistence".

DNA EXTRACTION, POLYMERASE CHAIN REACTION, AND SEQUENCING

DNA was extracted from leaf tissues of seedlings germinated from seeds grown in a greenhouse. The amplified region of sh4 included the entire first exon (847 bp) and a partial intron (153 bp) of sh4 gene and a 5' upstream flanking region (1509 bp) (Zhang et al. 2009). The DNA amplifying system followed that of Zhang et al. (2009). The internal sequencing primers were designed using the Primer3Plus, based on rice genome sequences. For the obligately outcrossing wild rice, the purified polymerase chain reaction (PCR) products were cloned into pMD18-T vectors (Takara, Dalian, China); eight randomly selected clones per seedling were sequenced. For highly selfing weedy and cultivated rice, the purified PCR products were directly sequenced in both directions. Singletons and ambiguous sites were re-sequenced to assure the quality of sequences used for analyses. All DNA sequences obtained for this study were deposited in GenBank.

SEQUENCE ANALYSES

Sequences of each rice accession were assembled by Seqman II (DNASTAR, Madison, Wisconsin), and the ends of amplicons were trimmed to remove low-quality sequences. Sequence alignments were performed using ClustalX and were refined manually. All aligned sequences were imported into the DnaSP package to extract all the nucleotide polymorphic sites on which our various haplotypes were identified. The FNP sites ("G" vs. "T"), generally considered to be responsible for rice's seed shattering/persistence, were identified for all sequences and compared among wild, weedy, and cultivated rice accessions.

RESULTS

WILD, WEEDY, AND CULTIVATED RICE SEED-SHATTERING/ PERSISTENCE PHENOTYPES AND SH4 FNP

All 166 wild rice accessions examined from more than a dozen Asian countries had a seed-shattering phenotype; likewise all 222 geographically diverse weedy rice accessions from many countries of three continents had the same seed-shattering phenotype. We examined the DNA sequences of all the 166 accessions of wild rice and 222 accessions of weedy rice from a wide geographic range to determine whether the FNP site of sh4 locus was the wild-type (putatively, seed shattering) "G" nucleotide or the mutational type (putatively, nonshattering) "T" nucleotide. We found considerable polymorphism in the wild rice samples. Almost three-quarters (73.5%) of the wild rice accessions contained the wild-type "G" nucleotide at the FNP site, whereas a sizeable minority (26.5%) of the wild accessions had the nonshattering "T" nucleotide. Interestingly, all (100%) weedy rice accessions examined had the inappropriate "T" nucleotide, although every weedy rice accession showed the seed-shattering phenotype. Thus, we found no relationship between presence of FNP "G" nucleotide or the "T" nucleotide and the seed-shattering/persistent phenotypes of wild and weedy rice. All 192 cultivated rice accessions, including both indica and japonica, examined were found to have the seed persistence phenotype.The sequencing data showed that all cultivated rice accessions examined had the "T" nucleotide at sh4's FNP.

SH4 LOCUS AND FLANKING REGION SEQUENCE POLYMORPHISMS IN WILD, WEEDY, AND CULTIVATED RICE

We compared the sequences of the entire first exon of sh4 and about 1500 bp of its upstream flanking region in our wild, weedy, and cultivated rice accessions. We found significant sequence polymorphism in wild rice accessions, where a total of 101 variable sites were identified. Based on these variable sites, we determined a total of 103 haplotypes for 166 wild rice accessions. In contrast, much less polymorphism was detected in the target sequences of weedy and cultivated rice accessions, with only one to three variable sites identified. We determined four haplotypes for weedy rice accessions and only two haplotypes for cultivated rice accessions (including indica and japonica rice). The haplotype frequencies in wild, weedy, and cultivated rice, as well as haplotype origins and associated shattering phenotypes.

For wild rice, nine "common" haplotypes, each represented by more than three accessions, accounted for 32.5% of all wild rice accessions. The 94 "rare" haplotypes, those represented by only 1–2 accessions, accounted for the rest, 67.5% of wild rice accessions. Of the nine "common" haplotypes, about 19% of the total wild rice accessions were identified to as haplotype-1 (H1) or haplotype-2 (H2) differing from one another by a single nucleotide substitution ("G" to "A") at 1158 bp site of the flanking region. About 14% wild rice accessions were the haplotype-3 to -9 (H3–H9).

In the 222 weedy rice accessions, only four haplotypes (H1, H2, H104, and H105) were identified. About 87% of the total weedy rice accessions were identified to be H1 or H2. Two weedy haplotypes were unique to specific regions; 24 weedy rice accessions from Italy and Spain were identified as H104, and four weedy accessions from the USA had H105.

In the 192 cultivated rice accessions, only two haplotypes (H1 and H2) were identified. H1 was represented by the majority (83.3%) of cultivated accessions. H2 was present in the minority (16.7%).

Thus, H1 and H2 are not only shared by wild rice, weedy rice, and cultivated rice; they are also the most prominent haplotypes for each. Interestingly, although all of the wild and weedy rice accessions were seed shattering, and all cultivated rice accessions were seed persistent, all wild, weedy, and cultivated accessions of H1 and H2 carried the mutational type of the "T" nucleotide at the FNP site of sh4, regardless of their seed dispersal phenotype. All other wild rice accessions (i.e., with haplotypes other than H1 or H2) carried the wild-type "G" nucleotide at the FNP site of sh4. The weedy rice accessions of H104 and H105 carried the mutational type "T" nucleotide at the FNP site, even though these accessions were seed shattering.

DISCUSSION

Our data demonstrate that some seed-shattering O. rufipogon and all seed-shattering weedy rice accessions have the "nonshattering" genotype with the "T" nucleotide at the FNP site of sh4, which challenges the widely accepted hypothesis that human selection for the "T" nucleotide (and selection against the "G" nucleotide) at the sh4 locus was a significant early step in the process of the domestication of Asian rice. In our large and diverse samples including those from published data (Li et al.2006; Zhang et al. 2009; Thurber et al. 2010), we found considerable discordance between Oryza seed-shattering/persistence phenotypes and putative "nonshattering" "T" and "shattering" "G" genotypes. Specifically, we found the "T" genotype at a remarkably high frequency (nearly 27%) in shattering wild rice (O. rufipogon complex) accessions collected over a wide geographic area.

While our screening is the numerically largest and geographically broadest to date, it supplements previous reports in which some wild rice (O. rufipogon) samples were also found to contain "T" nucleotide at the of sh4 locus. In fact, Li et al. (2006) who first reported the data suggesting a key role for sh4 in rice domestication also noted three accessions of nonshattering wild rice that contained the "T" nucleotide at the FNP site, the same as cultivated rice.

The authors' explanation was that these wild rice samples might be misidentified nonshattering weedy (not wild) rice that acquired the sh4 allele from its cultivated rice ancestor. Subsequently, Zhang et al. (2009) found two wild rice samples heterozygous for the "T" and "G" nucleotides at the FNP site of sh4, which they explained as "most likely the result of introgression of the nonshattering allele from cultivars into the wild populations." Our results from a large and geographically diverse set of samples are in accord with these previous reports of sh4 locus "G"/"T" polymorphism in wild rice populations. The so-called wild-type allele is far from fixation; the "T" allele is relatively common in wild rice populations. Thus, our results do not support the sh4 hypothesis because our large and geographically diverse set of wild rice samples we screened had a seed-shattering phenotype, independent of whether they were homozygous for either type of sh4 allele.

Interestingly, the sequence data for an even larger number and geographically more diverse of weedy rice accessions revealed that they were fixed for one allele, the "nonshattering" "T"-type of sh4 allele, despite the fact that all of those weedy rice plants had strongly seed-shattering panicles! Thurber et al. (2010) reported the same conclusion for a few dozen weedy rice accessions in the United States. All had seed-shattering panicles, but all had the "inappropriate" allele. Furthermore, Thurber et al. (2010) also found eight seed-shattering individuals of the O. rufipogon complex with the nonshattering "T" allele. They made a conclusion, "the presence of this mutation alone is not sufficient to confer a reduction in shattering". We agree. Collectively, these results strongly indicate that the presence or absence of wild "G"-type homozygosity of the sh4 allele does not affect the seed-shattering phenotype of wild and weedy rice. In other words, the data do not support an important role for the sh4 locus in the evolutionary shift from seed shattering to seed persistence during the domestication of cultivated rice from its wild ancestor.

Interestingly, about 19% of the phenotypically shattering wild rice accessions have the same H1 and H2 haplotypes (both having the "T"-type sh4 variant) as all cultivated rice accessions included in our study. These results suggest that cultivated rice was most likely domesticated from wild rice sharing the H1 and H2 haplotypes. Thus, one or more loci other than sh4 must have first conferred the shattering/persistent trait during the domestication of cultivated rice. Alternatively, introgression may also produce the same H1 and H2 haplotypes of wild and cultivated rice. However, the frequency of wild-crop introgression or recombination within such a small piece of DNA fragment (<1000 bp) within the sh4 locus should be extremely low.

In conclusion, our data and those from prior studies challenge the widely accepted hypothesis that the evolutionary replacement of the wild type of "G" allele by the "T" allele at FNP site of sh4 in wild rice populations was a substrate for human-favored selection for seed persistence as the major driving force for rice domestication. The considerable polymorphism we observed for the "G" allele and "T" allele in the seed-shattering wild rice plants from different populations does not support that hypothesis. In addition, the detection of the "T" allele in a large number of diverse accessions of seed-shattering weedy rice plants also rejects that conclusion.

We hypothesize instead that the sh4 "T" allele may have evolved to fixation in cultivated rice after the first steps of rice domestication in a role as modifier allele to an as-yet unidentified anti-shattering allele or alleles fixed at one or more other loci. Our hypothesis is consistent with the fact that the sh4 "T" allele co-segregates with nonshattering to a variable extent in certain QTL progenies, that the locus is expressed in the appropriate abscission tissues, and that cultivated rice transformed with the "G" allele has some increase in shattering (Li et al. 2006). Recently, Zhou et al. (2012) identified a seed shattering abortion1 (shat1) mutant in a wild rice introgression line (SL4).

This gene is proven to encode a transcription factor that affects seed abscission zone development, and therefore, also determines rice seed shattering. After comparing the DNA sequences of wild and cultivated rice at this locus (sequences data from GenBank), we found no sequence polymorphism in the examined sequences as described by Zhou et al. (2012). We think that shat1 should not be the unidentified locus responsible for the domestication of cultivated rice either. Further research is necessary to identify the original loci and alleles responsible for the initial reduction in shattering during the first steps of rice domestication.

Thus, we hypothesize that an unidentified locus or loci are responsible for the domestication of cultivated rice through reduced seed shattering. Once the alleles were fixed, modifier alleles at other loci were selected to further decrease seed shattering. The "T" allele at the FNP site of sh4 is likely to be one of these that were incorporated late in the domestication process. We hypothesize that decreased shattering expression due to the "T" allele can only occur in the presence of an appropriate genotype or genotypes at other loci that were previously selected during domestication. The allele at qsh1, mentioned in the Introduction, is another example of an allele that further reduces seed shattering and appears to have been selected by mechanical threshing long after domestication during the crop improvement phase of temperate japonica.

The evolutionary process of rice domestication is filled with mysteries yet to be examined. Identifying the unknown alleles associated the origin and genetic mechanisms of domestication is important not only for cultivated rice but also for understanding those processes in other crops. Likewise, that information has practical applications, including the proper management of weedy rice whose persistence in rice fields through seed shattering causes significant rice yield losses.

For example, silencing the allele-based seed-shattering expression using RNAi biotechnology as suggested by Gressel and Valverde (2009) may mitigate the agronomic or ecological impacts caused by transgene flow to wild and weedy rice populations. Silencing the expression of the true shattering allele(s) should significantly reduce seed shattering in weedy plants that have acquired transgenes by gene flow, and consequently reduce their seed dispersal and fitness (Gressel and Valverde 2009).

GENOMIC PATTERNS OF NUCLEOTIDE DIVERSITY IN DIVERGENT POPULATIONS OF U.S. WEEDY RICE

Among the most widespread and costly agricultural pests are the numerous weeds that have evolved from within the same complex of interfertile species as domesticated plants. The recent and rapid evolution of these conspecific weeds also presents unique opportunities to study processes influencing adaptive population divergence and parallel evolution of weedy life-histories. Conspecific weeds are morphologically and ecologically divergent from domesticated and wild congener species, and are not simply transient "volunteers" of the previous season's crop. The evolutionary success of conspecific weeds is often attributed to acquisition of traits associated with wild plants (e.g. dormancy), presumably selected against in crops. Conversely, these weeds also often exhibit characteristics typical of domesticated plants, (e.g. more selfing, rapid growth), which could promote invasiveness in the agroecosystem. There is great interest in understanding the evolutionary mechanisms that can lead to the emergence of weedy species from the same species complexes that give rise to domesticated plants.

The larger complex of interfertile species within which conspecific weeds evolve includes the crop, wild relatives, and other feral weeds. Studies have shown that, in many cases, hybridization between crops and wild species can facilitate weed evolution [reviewed in]. Alternatively, conspecific weeds may evolve from standing genetic variation in wild relatives, or cultivated germplasm, though examples of weeds evolving directly from crops are rare. The short evolutionary time scales involved make it less likely that novel mutations are significant to weed evolution, however exceptions are known.

Here we investigate the evolutionary origins of weedy rice in the United States, which has been a subject of considerable debate for more than 150 years. Weedy or red rice (due to the frequent presence of a red pericarp), is found in cultivated rice fields worldwide, but is most damaging in direct seeded (seeding directly into a dry soil bed), highly mechanized agricultural systems typical of the U.S., Europe and Australia. Although currently classified as the same species as Asian cultivated rice, Oryza sativa L., weedy rice has morphological characteristics typical of wild species (e.g. dormancy, shattering) and of cultivated rice (e.g. high fecundity, high selfing rate). The long term persistence of weedy rice throughout the range of cultivated rice, suggests that it can adapt to local changes in agronomic practices as well as different biotic and abiotic conditions.

No Oryza is native to the U.S.; therefore, U.S. weedy rice must have evolved elsewhere and/or endogenously from introduced cultivated and/or wild germplasm.

The Oryza crop-wild complex, within which weedy rice evolved, is composed of two domesticated and six wild species that share the AA genome. Evidence for gene flow among members of this complex is extensive, suggesting that any of these taxa could have contributed to the origins of weedy rice in the U.S. The earliest available reference to weedy rice in the U.S. dates from 1846, and describes a well-established and troublesome pest. Considerable phenotypic diversity is found within U.S. weedy rice populations. Currently, two main morphological groups include awnless straw-hulled types, which more closely resemble cultivated rice varieties, and awned black-hulled forms, with other morphologies found less frequently. Several SSR and RAPD studies have suggested that strawhull, awnless weedy rice is most closely related to indica, O. sativa varieties typical of lowland tropical regions, and probably a product of hybridization with O. rufipogon/O. nivara, the wild ancestor of domesticated Asian rice. A recent microsatellite study suggests that some black hull, awned weedy rice may be derived from O. sativa aus varieties, a group most commonly grown in Bangladesh and Northeastern India, or from O. rufipogon.

To date, however, patterns of DNA sequence diversity have not been explored in U.S. weedy rice, and open questions remain about the likelihood of weed origins from cultivated ancestors, and the roles of demographic history and hybridization in the evolution of weedy rice.

Taking advantage of the existing genomic resources for domesticated rice, we use genome-wide patterns of DNA sequence variation in a broad sample of the Oryza crop-wild complex, to infer the origin and demographic history of U.S. weedy rice. Specifically we attempt to address remaining uncertainties regarding 1) the ancestral Oryza group(s), including other wild species, that gave rise to U.S. weedy rice, 2) the timing of divergence between U.S. weedy rice and its progenitor(s), and, 3) the role of hybridization in the establishment of U.S. weedy rice populations. We find considerable population structure in U.S. weedy rice, with genetically divergent populations having separate origins. Exotic cultivated O. sativa varieties are the main contributors to weedy rice genomes, and there is little evidence of contribution from wild Oryza. Hybridization among weedy groups has also influenced the emergence of novel weed phenotypes.

Assessments of demographic parameters suggest differences among divergent weedy groups in the effect of population bottlenecks upon U.S. colonization, and in the timing of their origins. Our results demonstrate how similar weedy life histories can evolve from divergent genetic backgrounds.

METHODS

PLANT MATERIAL

Weedy rice

Weedy rice seed was obtained from collections made over a period of 30 years in the Southern rice belt (Arkansas, Louisiana, Mississippi, Missouri and Texas) and maintained by the United States Department of Agriculture (USDA) at the Dale Bumpers Rice Research Institute, Stuttgart Arkansas. We selected a subset of 58 accessions that maximized geographical diversity, but were otherwise chosen at random. We also included a few samples representative of rare morphologies (i.e. brown hulls), to increase the probability of capturing all existing population structure.

Putative parental populations

For our analyses, we used data from 206 Oryza accessions, 95 of which were included in, and 111 which were chosen specifically for this study. Our sample broadly surveys AA genome Oryzaspecies for potential parental sources of U.S. weedy rice.

We included Asian landraces and modern accessions from the five main variety groups of O. sativa; this includes 22indica, 7 aus, 18 tropical japonica (varieties grown in tropical and subtropical regions), 22temperate japonica (varieties typical of northern latitudes), and 6 aromatic (fragrant rice varieties). A plurality of evidence supports the independent domestication of the indica and ausgroups from the japonica and aromatic groups beginning ~ 10,000 ybp from divergent populations of O. rufipogon. An additional 12 tropical japonicacultivars were added that are representative of important U.S. founding lineages (i.e. Carolina Gold, Blue Rose;) or have been extensively grown in the southern U.S.

We included 50 O. rufipogon and 3 O. nivara (a species often considered an annual form of O. rufipogon) accessions, sampled across their geographic range.

More samples from India and China were included as these regions are the possible centers of origin for domesticated rice. Four accessions of African domesticated rice (O. glaberrima) and three of its wild progenitor (O. barthii) were included, as historical evidence suggests their introduction by early crop breeders and Africans brought to the U.S. as slaves. Similarly, two accessions of O. glumaepatula were included, as it occurs in the Caribbean and Central America, and may have contributed to the evolution of weedy rice. O. meridionalis, native to Australia and Oceania, was included as an outgroup, as phylogenetic evidence indicates that it is ancestral to other AA genome Oryza.

DNA extraction and sequencing

DNA was extracted from approximately 1 g of fresh leaf material from one plant per accession using a modified CTAB protocol. DNA concentrations were gel quantified and diluted to 2 ng/ul for sequencing. We amplified and sequenced a total of 48, ~ 400-600 bp, gene fragments, selected from a set of 111 randomly chosen sequenced tagged loci (STS) developed by. The 48 fragments were chosen to include ~4 loci per chromosome distributed on both chromosome arms, without referencing diversity data or estimates of informativeness.

DNA sequencing was carried out in Cogenics sequencing facilities (Houston, TX) as described in. Base pair calls, quality score assignment and construction of contigs were carried out as described in. Newly constructed contigs were added to existing alignments, and all subsequent analyses were based on the merged alignments. Further sequence alignment and editing were carried out with BioLign Version 2.09.1 (Tom Hall, NC State Univ.) as described in. New DNA sequences obtained for this study were deposited in GenBank under accession numbers GQ999668-GQ999777.

The cytoplasm genomes of O. sativa cultivars from independent domestication events have been used to distinguish cultivar groups. We assessed the origins of cytoplasm genomes in weedy rice using one chloroplast, and two mitochondrial [SSV500 and SSV39,] markers in all 58 weedy rice accessions, and 82 Oryza samples from our panel and those from for which DNA was available. These PCR-based markers amplify regions in the chloroplast or mitochondria containing large indels (69 bp to 500 bp), which can be visualized on a 1% agarose gel. We assumed maternal inheritance for cytoplasmic genomes, and combined the three markers into a single cytotype for analysis.

Population structure

We assessed population structure using the Bayesian clustering program InStruct, which is similar to the commonly used STRUCTURE, but was developed specifically for identifying population structure in inbreeding species. Cultivated and weedy Oryza tend to self-fertilize, while wild Oryza outcross more frequently (10 to 60%). InStruct does not assume Hardy Weinberg equilibrium within populations, which can result in over-splitting in populations with a history of inbreeding. We created genotype data from phased haplotypes inferred for each STS fragment using PHASE 2.1.

We inferred population structure using two data sets: one included only U.S. weedy rice accessions (N = 58) and the second contained all individuals used in this study (N = 209). To determine the number of populations (K) that best approximates population structure, we tested a range of purposefully extreme K: K = 2 to 20 for the complete data set, and K = 2 to 15 for the weedy rice dataset. For each value of K, five replicates were carried out with an initial burn-in of 100,000 followed by 500,000 iterations using the "infer population structure and the individual selfing rates" option for final simulations. Sizes of burn-in and simulation number were found sufficient based on the Gelman-Rubin estimate of chain convergence for preliminary trial runs of various lengths (data not shown). All InStruct analyses were run on a computer cluster freely available at the Computational Biology Service Unit of Cornell Universityhttp://cbsuapps.tc.cornell.edu/InStruct.aspx webcite. We used the Deviance Information Criterion (DIC) scores provided in the InStruct output to determine the number of populations that best fit our data. The K with the lowest average DIC score of the five replicates was considered to best describe population structure. For the model with the lowest mean DIC score, we checked for consistency in estimates of membership coefficients and split locations by estimating the correlation between ancestry membership matrices of replicate model runs with the R package simco. InStruct results were plotted using R v2.6.2.

Summary statistics

Summary statistics for each STS locus and population of interest, including nucleotide diversity (?Wand ?p), Tajima's D, polymorphic loci (P), number of segregating sites (S), and population unique alleles/haplotypes were calculated as described in. Site type determination was based on annotations of the O. sativa genome (TIGR v. 5 January, 2008).

Levels of population differentiation were estimated using Fst, calculated after, using modifications of, which drops singleton SNPs. We calculated Fst for each STS fragment by taking the mean Fst of all SNPs per fragment, and then calculated the grand mean over all STS fragments, counting non-polymorphic fragments as zeros. Negative values of SNP Fst were changed to zero before taking means of individual SNPs per STS fragment.

DEMOGRAPHIC MODELS OF WEEDY RICE EVOLUTION

To infer the demographic history most consistent with the observed patterns of polymorphism in U.S. weedy rice, we used a full likelihood method, IMa (Isolation with Migration analytic;), and an approximate Bayesian computation (ABC) method that relies on summary statistics.

A description of the demographic model and assumptions of the IMa analyses are provided in Three population pairs were considered, and each IMa analysis used only STS polymorphic within each population pair, as preliminary runs including invariant loci would not converge in a reasonable time. All pairs contained a similar number of polymorphic loci (27-32); thus exclusion of invariant loci does not preferentially affect parameter estimates in any group. We used a neutral mutation rate of 1 × 10-8, derived from synonymous site divergence at the maize Adh loci to convert ML estimates to years and number of individuals. Both cultivated and weedy rice are, on average, annual plants under field conditions due to harvesting and cultivation practices, and we assumed a generation time of one year. Note that excluding monomorphic STS effectively increases the baseline mutation rate by ~1.6 (48/30), but this value is within error ranges of mutation range estimates, and does not affect scaling of parameters across groups. For all runs, we assumed that migration between populations was symmetrical, and set the maximum prior for population sizes to be equal. For final runs, we used a burn-in of 5,000,000 and recorded simulations for an additional 5,000,000 iterations using 10 chains and a two-step geometric heating scheme. To check for convergence, we ran each parameter set three times with a different starting random seed. IMa command lines were: ima -b 5000000 -l 50000 -m1 25 -m2 25 -f g -n 10 -g1 0.7 -g2 0.8 -p345 -q1 5 -k 3 -t 5 -s12307.

The demographic model used in ABC analyses, and is similar to models used to assess population divergence and crop domestication. We used this model to test which scenario is most consistent with the demographic history of U.S. weedy rice: i) prior to introduction to the U.S., from a domesticated progenitor in Asia (~12,000 years before present); ii) de novo, from cultivated germplasm introduced to the U.S.; and/or iii) from wild populations prior to domestication All simulations were performed using MS, and were conditioned on the population mutation rate ? = 4Nμ, where 4N is the reference population size and μ is the per nucleotide per generation mutation rate (μ = 1 × 10-8, as above). We used the observed mean silent site ?w for O. rufipogon from to estimate ? = 4Nμ, as the allele frequency spectrum of O. rufipogon is consistent with a population evolving at a constant size. The population recombination rate, ?, was assumed to be identical to ?, similar to other recent studies. We considered weedy rice to be effectively entirely selfing and scaled timing parameters using 2N, rather than 4N.

We assumed that the population size of the progenitor of weedy rice has remained constant and set ?c equal to ?p for the duration of an individual simulation. Priors for ?c were based on the ratio ?c/4N and ranged from 0.1 to 0.7. These limits were based on the observed ratio of silent site ?w, crop/ ?w,O. rufipogon. Priors for the current and bottleneck population size of the weed were based on the ratios ?r/?c, and ?b/?r, and ranged from 0 - 1 and 0-?r respectively.

Priors for time of population expansion (tg), founding in the U.S. (tf), and time of divergence (ts) were based on the known history of cultivated rice in the U.S. and timing of domestication. The upper limit for tg was chosen to coincide with the rapid expansion of cultivated rice, which began around 1870 in the southern rice belt. Similarly, tf was assumed to have occurred after cultivated rice was introduced into the U.S. and was constrained to be less than 400 years ago. Priors for ts ranged from tf to 50,000 ybp, and were chosen to be consistent with divergence occurring prior to domestication (ts= 12,000-50,000), post domestication in Asia (ts = 12,000), and at the time of founding in the U.S. (ts = tf). A grid of prior values for the three timing parameters and ?c was generated, and the MS command line and further details on parameter ranges.

Summary statistics and observed data were calculated using data pooled from all 48 STS fragments. We chose summary statistics shown to be sensitive (correlated) to changes in population growth and timing of divergence. These statistics also illustrate a key pattern observed in the data: that weedy rice groups contained a subset of genetic diversity present in putative ancestral populations. We used eight statistics: ?p for both populations combined, the number of segregating, fixed, and private sites in weedy populations and their putative cultivated progenitors, and the number of shared sites between weeds and their putative progenitors. A similar set of summary statistics were used to infer demographic history in Zea.

We employed a similar rejection approach as in and used the proportion of accepted simulations to calculate the approximate likelihood for a given demographic scenario. For each of the scenarios described above, we performed ~850,000 simulations. All processing and analysis of MS output were performed using R.

RESULTS

Marker data

The 48 sequenced STS ranged in aligned length from 400 to 921 base pairs (bp) over all accessions, for a total of ~24,145 bp aligned sequence per accession. We observed 827 SNPs in our entire dataset. Thirty-three SNPs had more than two alleles, primarily (73%) due to alternative states present in the outgroup species (O. barthii or O. meridionalis). These SNPs were excluded from analyses when occurring in targeted groups. Insertions and deletions (indels) were not used in haplotype determination or calculation of summary statistics (unless segregating sites occurred within an indel, which was rare). Heterozygotes were observed almost exclusively in O. rufipogon, and only two weedy rice and four cultivated O. sativa accessions had heterozygous sites.

Except for one weedy rice accession, the three-cytoplasm markers amplified in all individuals screened (n = 139). We observed the same sized length variants (i.e. the size of deletion in base pairs) for each marker that were found. In general, we found that the cytotypes had similar distributions within cultivated O. sativa varieties as reported in and. However, unlike, we did not find complete linkage between the mitochondrial markers.

U.S. weedy rice population structure

To determine the number of weedy rice populations occurring within the U.S., we used InStruct and a data set containing only weed accessions. Based on DIC scores, we found that population structure is most consistent with a model containing six groups (K = 6). Individuals belonging to the same cluster tend to have similar grain morphologies. At K = 2, individuals with straw hulls that lack awns (SH = straw hull) are differentiated from other hull phenotypes.

With increasing K, SH individuals remained in a single cluster, while the non-SH group was further subdivided into five subpopulations. Based on the predominant grain phenotype (i.e. hull color and presence or absence of an awn) in each population, we designated these as: BHA1 (black hull awned 1), BHA2 (black hull awned 2), BRH (brown hull awned), MXSH (mixed straw hull), and MXBH (mixed black hull awned). With one exception, all 24 weedy rice individuals with straw-colored hulls and no awns in our panel clustered in the SH population. All other clusters, however, contained multiple grain phenotypes. For example, ~73% of individuals in BHA1 and ~63% of BHA2 had awns and a black hull, and ~60% of BRH individuals had brown hulls and awns. Similar results were obtained when analyses were run with STRUCTURE (data not shown).

Oryza population structure

To identify potential source(s) for U.S. weedy rice within Oryza, we used InStruct and a dataset that included all accessions in our panel (n = 209). The best fitting model contained nine populations (K = 9). Cluster membership was generally consistent with previous research. InStruct identified O. sativa varieties aus, indica,tropical japonica and temperate japonica as distinct populations; however, our dataset lacked resolution to differentiate tropical japonica and aromatic accessions. The fourteen U.S. cultivars included in this study clustered with tropical japonica, as expected, and historic and modern cultivars were not differentiated.

Approximately four clusters were observed within the wild ancestor of cultivated rice, O. rufipogonalthough most individuals appeared to be admixtures. Many O. rufipogon individuals shared some ancestry with indica, but only five had membership coefficients greater than 50%. None of these were indicated as hybrids in the passport data available, and admixture may be due to shared ancestry, rather than recent hybridization. Consistent with previous research no distinct O. nivara cluster separate from O. rufipogon was observed. African cultivated rice, O. glaberrima, and its progenitor, O. barthii, formed a distinct cluster, as did the two O. meridionalissamples. O. glumaepatula samples, on the other hand, clustered with three O. rufipogon and oneO. nivara.

Origins of U.S. weedy rice populations

To determine the putative progenitors of U.S. weedy rice, we used the results of the two InStruct analyses, combined with the genotyping results for the three-cytoplasm markers. All of the SH individuals identified by InStruct cluster with indica when all samples are used. All SH accessions had the same cytotype, which was also the most frequent in indica (60%) and O. rufipogon (53%), and was found in all of the O. rufipogon and O. nivaraaccessions that shared greater than 50% membership with indica.

Both black hulled weedy rice groups, BHA1 and BHA2, cluster primarily with aus and are not differentiated in the InStruct analysis that included all individuals. Interestingly, the most frequent BHA1 (60%) and BHA2 (71%) cytotype did not occur in our aus sample, but is most common in tropical japonica (63%), and rare in indica (20%) and O. rufipogon (7%). However, two other cytotypes found in BHA1 and BHA2 were also found at high frequency in aus. Two BHA1 individuals and an O. rufipogon accession from India shared a cytotype that was absent in all other accessions.

The InStruct analyses suggest that the BRH population is either the result of hybridization betweenindica and aus, the SH and BHA weedy groups, indica and BHA, or aus and SH. The BRH group contained a subset of the diversity found in the SH and BHA groups (10 of the most frequent STS haplotypes [MFH] in BRH were exclusive to BHA1 and BHA2, and six to SH; the remaining 32 were common to all weedy populations) consistent with hybridization among weedy groups in the U.S. All BRH individuals have the same cytotype as SH weeds, suggesting a maternal SH lineage. No heterozygotes were observed, which would be expected from early generation hybrids; however, heterozygosity may have been affected by selfing at the USDA stock center.

InStruct results also indicate that hybridization between tropical japonica varieties grown in the U.S. and weedy rice has occurred. Population MXSH contains two individuals that share genetic membership with both indica/SH and tropical japonica. The MXSH population is also notable in that weedy rice is likely the paternal rather than maternal parent, as observed cytotypes are absent from SH weeds, but occur in tropical japonica. Individuals in the MXBH group were identified as admixtures between aus/BHA and tropical japonica. Both accessions in MXBH have the same cytotype, which is absent in aus, but found in BHA groups and tropical japonica. Outside of the MXSH and MXBH populations, only one accession shared membership with tropical japonica.

Three of the five putative hybrids we identified were listed as suspected crosses based on morphological observations made at the time of collection. Three modern U.S. cultivars (M202, Bengal, and Palmyra) appear as admixtures of temperate and tropical japonica in our analyses, in agreement with known pedigree data. This suggests our data is sufficient for identifying relatively advanced generation hybrids and supports our designation of weedy hybrids.

Genetic diversity in weedy rice

Genetic diversity statistics were calculated for all 48 STS for U.S. weedy rice groups and potential sources within Oryza. For all measures of genetic diversity, weedy rice, considered as a single group, was genetically depauperate, with less than ~40% and ~70% of the diversity present in O. sativa and O. rufipogon respectively. Mean ?W and ?p are highest in O. rufipogon, intermediate for O. sativa groups and lowest in U.S. weedy rice. We also found that the distribution of nucleotide diversity is heterogeneous across loci for all groups, but particularly in weedy and cultivated groups where a few loci are atypically polymorphic.

In general, U.S. weedy rice groups contain a subset of diversity observed in their most closely related cultivated O. sativa populations. SH weedy rice contains only ~30% of the silent site variation found in indica, while the BHA1 and BHA2 groups harbor between 50-67% of the variation found in aus. For a majority of the STS fragments, weedy rice groups and putative progenitor shared the same MFH (83% of STS fragments in indica and 73% of STS fragments in aus). We did not observe high frequency or specific haplotypes that would suggest weedy rice is a product of recent hybridization with O. rufipogon.

We estimated differentiation between U.S. weedy rice populations and their closely related Oryzagroups using mean and median values of STS Fst. SH and BHA1 were more diverged than indica from aus, and the most highly differentiated population pair tested. Populations BHA1 and BHA2 were not greatly differentiated but both were more diverged from austhan SH from indica, consistent with a higher number of private SNPs and presence of one fixed SNP in BHA/aus comparisons. Low median Fst values between weedy populations and their closest Oryza relatives show that estimates of population differentiation are driven by a few loci.

ESTIMATES OF DEMOGRAPHIC PARAMETERS

We used IMa and ABC to infer time of divergence and population sizes for the two main weedy rice groups (SH and BHA1) and their closest Oryza relatives. The results of a single simulation for demographic parameters from IMa analyses, and posterior probability density curves for parameter estimates. For all population pair comparisons, differences in parameter estimates among the three simulation runs were small (less than 5%) and the 90% posterior density intervals (HPD) overlapped, suggesting chain convergence. The maximum likelihood (ML) estimates of current and ancestral effective population size (Ne) were consistent with expectations that U.S. weedy rice populations have experienced population bottlenecks. ML estimates of Ne for BHA1 (~2,472 individuals) and SH (~1,000 individuals) are an order of magnitude smaller than for their ancestral populations (77,148 and 74,397 respectively). ML estimates of current weedy rice Newere also smaller than estimates of current Ne for their relatives. Ne estimates for aus and indicaacross simulations were similar, with HPD intervals covering similar ranges in population size. The larger estimates for indica are consistent with the limited geographical distribution of aus. To account for high rate of selfing in cultivated rice, our analysis is based on a single haplotype (i.e. chromosome) per individual, and, therefore, depending on the actual degree of selfing (exact values are unknown), population size estimates may be, at most, twice as large.

IMa-based estimate of divergence time between aus and indica was ~6,047 ybp,, with a wide HPD interval (605 to 241,880 ybp). Divergence time estimates for SH from indica (~31,995 ybp) and BHA1 from aus (~9,939 ybp) predate the introduction of cultivated rice to the U.S. (~1690's), and its establishment in the southern rice belt (>150 years). However, confidence HPD intervals for all estimates are very large and overlap.

Obtaining estimates of migration between populations from our IMa runs was problematic. Initial runs of models that did not include migration, under the assumption that gene flow between weeds in the U.S. and cultivars in Asia is unlikely, did not converge. Including migration improved estimates for remaining parameters. However, for all population pairs, estimates of migration are not reliable, as posterior distributions did not converge within prior ranges, suggesting that, under short evolutionary time scales, with this dataset, IMa may confound recent divergence with ongoing gene flow.

To further explore the demographic history of U.S. weedy rice, we carried out coalescent simulations under four demographic scenarios, and compared obtained summary statistics with those of the observed data. For both weedy rice groups, divergence from cultivated relatives prior to domestication was not supported.

However, divergence prior to arrival to the U.S. was supported for BHA1 from aus. For SH weedy rice, recent divergence from indica, occurring either in Asia or in the U.S. was found to be most likely (Table5). Similar to our IMa analyses, we found that population bottlenecks have played a role in the evolution of U.S. weedy rice, but bottleneck intensity appears to have impacted SH more than BHA1. For both weedy populations, the distribution of tg indicates recent and simultaneous population expansion, consistent with the known history of expansion of cultivated rice production in the past 100 years in the Southern rice belt. Founding in the U.S. for both BHA1 and SH appears to have occurred within the past 200 years.

DISCUSSION

The evolutionary origins of U.S. weedy rice

Current weedy rice populations in the U.S. are morphologically diverse, and we find that population structure in weedy rice is correlated with hull morphology. The two major weedy rice groups occurring in the U.S. are most closely related to the exotic cultivated rice varieties, aus and indica. Our data thus provides strong evidence that weedy con-generics can evolve directly from domesticated backgrounds, a result that has been little reported/confirmed to date.

Similar to previous morphometric and molecular marker studies, weedy rice individuals that have straw hulls and no awns (SH) cluster primarily with O. sativa indica. Other hull morphologies, including black and straw hull with awns (BHA1, BHA2), cluster primarily with O. sativa aus, a relationship also recently detected with microsatellites. Unlike previous microsatellite based studies, we did not find conclusive evidence for contribution of wild Oryza species to U.S. weedy rice. Although some O. rufipogon and O. nivara accessions clustered with indica and SH weedy rice, only two out of 51 accessions had the same level of shared genetic membership (>80%) with SH weedy rice as all indica accessions. Moreover, accessions of O. rufipogon and O. nivara that clustered with SH groups in our analysis do not share hull morphology or any unique alleles with weedy accessions, unlike indica, supporting shared ancestry as the most likely explanation for the clustering pattern. Black hulls and awns are a common phenotype in O. rufipogon and O. nivara, but no accessions from this group clustered with BHA weedy rice. Aus cultivars, however, often have dark hulls and awns. The results of our clustering analyses combined with morphological data suggest that the main U.S. weedy rice groups evolved primarily from aus and indica genetic backgrounds.

Although clustering of weedy rice groups with cultivated relatives could also be due to common descent from a shared ancestral founding gene pool, the pattern of shared polymorphisms among weedy and cultivated groups is more consistent with direct descent from domesticated ancestors. Most of the SNPs found in the SH and BHA groups are a subset of those found in indica and aus, respectively.

This is particularly striking for the SH group, which contains only one non-singleton SNP not also found in indica, fewer than what it has with respect to O. rufipogon. Moreover, in each main weedy rice group (SH and BHA1), the most frequent haplotype (MFH) at each STS locus was most often the MFH observed in its putative progenitor group (data not shown). The greater divergence and number of private SNPs seen in the BHA groups with respect to aus, as well as differences in some cytotypes, however, suggest that demographic histories (e.g. magnitude of bottleneck, founding events, time of introduction) differ between the BHA and SH groups.

The close relationship of weedy rice with cultivated groups not grown in the U.S. suggests that both major weed groups were introduced either as stock seed contaminants or escaped breeding material. Although the majority of rice grown commercially in the U.S is tropical japonica, extensive opportunities for the intentional and unintentional introduction of Oryza germplasm have occurred. During the establishment of rice industry in the southern rice belt (~1860-1900), rice germplasm collected by the USDA was given to farmers directly for testing, potentially facilitating the spread and escape of weedy rice. During this time, farmers also commonly purchased seed from outside the U.S., which likely included representatives of all major O. sativavarieties.

The timing of weed evolution

If U.S. weedy rice groups originated from cultivated ancestors, it is of interest to determine whether divergence of the weeds occurred prior to or concurrent with their introduction to the U.S., and how divergence is related to the timing of domestication. We first estimated divergence time between aus and indica cultivar groups, which likely stem from the same domestication event. The ML estimate of ~6,000 years is reasonable, given that the commonly accepted time for domestication is ~10,000 years ago; however, confidence intervals for the estimate are large, consistent with the difficulty in estimating population parameters for very recent events. In contrast, IMa estimates for divergence of weedy groups from their cultivated relatives were surprisingly ancient, although, again, confidence intervals were very large. ABC coalescent simulations, on the other hand, supported a very recent SH-indica divergence, within the past 100 years, but divergence for BHA1 and aus occurring within the past 7,000 years.

We considered two possible explanations to account for the discrepancy between SH-indicadivergence time estimates obtained in each of our analyses. First, contribution of other groups to the weedy rice gene pool or unsampled variation in the putative progenitor could violate IMa assumptions that gene flow occurs only between population pairs, inflating estimates of divergence time. However, SH weedy rice contains a subset of the nuclear and cytoplasmic genetic diversity inindica; the only non-singleton private SNP in SH occurs at low frequency (13%), and was not found in any other Oryza group other than BHA weeds. Thus, introgression or incomplete sampling of indica diversity is an unlikely explanation of divergence estimates.

Alternatively, IMa divergence time estimates may be affected by the combination of an extremely strong bottleneck coupled with very recent divergence between indica and SH. Although, the IMa model is particularly suited to recently separated populations that are not under equilibrium, simulation studies to test sensitivity of IMa to extremely recent splits with no accumulation of divergent mutations have not been done (J. Hey, personal comm.). Both demographic analyses and the low levels of observed polymorphism support a very strong bottleneck for SH weedy rice. Since few lineages seem to have founded this weedy group, the divergence times obtained may represent the coalescence of these founders with the entire indicagene pool, and not the more recent split between weedy rice and progenitor. Observed patterns of polymorphism support the more recent divergence time estimated by ABC: SH either diverged from indica concurrent with its establishment in the U.S. (maximum of 400 ybp), or within 1000 ybp.

Both demographic analyses suggested an older divergence of BHA1 from its putative ausprogenitor, either after domestication, or close to the timing of domestication. In addition to one fixed site, the BHA1 group contains some private SNPs and cytotypes not observed in our aus sample. These patterns of polymorphism may indicate introgression of other Oryza, or incomplete sampling of aus diversity, which could have an effect on estimates of divergence time. Nuclear SNPs observed in BHA1 but not aus occurred at moderate frequencies (average 52%) and were also relatively frequent in other groups such as O. rufipogon, O. nivara, and tropical japonica, supporting the possibility of introgression. However, our Instruct analysis did not detect contribution of other Oryza groups to BHA weedy rice, and we have no a priori reason to believe our aus sampling did not capture the genetic diversity present in this geographically limited group.

Given the shared ancestry of all cultivars, weeds, and O. rufipogon, BHA1 private alleles shared with other groups could be a result of lineage sorting since divergence from aus. Interestingly, the single fixed SNP differentiating BHA groups from aus was not observed in any other Oryza group, supporting longer divergence between BHA weedy rice and its putative progenitor. Our estimates suggest that the founders of the BHA1 weedy rice group split from their cultivated relatives several thousand years ago and therefore may have existed as weeds prior to their introduction to the U.S. The ABC analysis marginally supported the introduction of BHA weeds before SH, which is contrary to expected based on historical records; black hull awned plants were not recorded until the 1920's, and anecdotal evidence attributes their origin to a cultivar introduced to Louisiana and abandoned due to excessive shattering

The role of hybridization in U.S. weedy rice

In addition to multiple introductions, our results suggest that hybridization and introgression occurring post-founding have contributed to the development of morphological diversity in weedy rice populations. The BRH population is most probably a product of hybridization occurring in the U.S. between SH and BHA weedy rice. No indica or aus are grown in the U.S., and therefore, an additional introduction of a weedy or cultivated group to the country would be required if BRH were the result of hybridization between indica-aus, indica-BHA, or SH-aus.

The high estimates of Fst between SH and BHA1 (~0.32) indicates that gene flow is relatively infrequent between weedy groups. Prior research has suggested non-overlapping flowering time, high selfing rates, and height differences as possible mechanisms restricting gene flow between straw-hull and black-hull awned weedy types.

Evidence that tropical japonica cultivars grown in the U.S. have contributed to genomic backgrounds of weeds in our sample set is limited to a few individuals in the MX populations. Several studies have observed both pre- and post-zygotic reproductive isolating barriers in experimental crosses between tropical japonica and weedy rice.

The existence of some barrier to gene flow is supported by the lack of more extensive hybridization in our sample. However, the barrier is "leaky," as both BHA and SH-tropical japonica hybrids are found. Additionally, the maternal lineage of at least one hybrid was consistent with weedy rice being the paternal parent, and therefore, gene flow from the weed to the crop could be an alternative pathway for weed evolution. Although infrequent, the fact that hybridization occurs at all presents a challenge to the management and continued use of cultivars containing traits suspected to increase weed fitness, such as herbicide resistance.

THE BIOLOGY OF ORYZA SATIVA (RICE)

SPECIES COMPOSITION AND DISTRIBUTION

Northern India, Southeast Asia, and southern China are believed to be the centre of origin of Asian rice (Oryza sativa). The rice genus Oryza has a pantropical distribution and comprises approximately 23 species that include both diploids (2n=2x=24) and tetraploids (2n=4x=48), and ten different genome types: AA, BB, CC, BBCC, CCDD, EE, FF, GG, JJHH, and JJKK (Vaughan 1994; Ge et al 1999). The genus Oryza is distributed in Asia (eg,O. rufipogon and O. nivara, both AA genomes), Africa (eg O. barthii, and O. longistaminata, both AA genomes), Australia (eg, O. meridionalis, AA genome), tropical America (O. glumaepatula, AA genome, and O. grandiglumis, O. alta, and O. latifolia, all CCDD genomes) (Akimoto 1998; Sano and Sano 1990; Vaughan 1994). Asian cultivated rice (O. sativa) has the diploid AA genome. Wild progenitors of African cultivated rice (O. glaberrima, AA genome) are grasses endemic to West Africa.

ENVIRONMENTAL SAFETY CONSIDERATIONS

Outcrossing and weediness potential

Cultivated rice (Oryza sativa L.) is primarily an autogamous, self-pollinating plant, although gene introgression into other cultivated rice is possible. Cultivated rice is an annual crop, it does not shatter or disperse its seed, and it has not acquired extended dormancy. Reported outcrossing rates are less than one percent and are limited by the biological characteristics of rice.

Factors including flower morphology, inability of pollen to remain viable longer than a few minutes, and a lack of insect vectors for pollen spread contribute to the low propensity of rice to cross-pollinate. Modern rice cultivars are often grown near older, traditional landraces in Asia, whereby only very low hybridisation rates between these two groups have been observed (Rong et al 2004). This is consistent with recommended distances of six metres, and even less in certified seed production (Gealy et al 2003). Oryza species with different genome types have significant reproductive isolation, making them unlikely to hybridise with each other. Hybridisation between species in different genera within the tribe Oryzeae is extremely difficult, even using artificial conditions, such as embryo rescue.

In the United States, the only wild species known to be compatible with cultivated rice is O. rufipogon, which has been found in a single location in the Everglades of Florida, and red rice, a wild variant of cultivated O. sativa, thus it is considered very unlikely that cultivated rice would hybridise with O. rufipogon under such conditions.

Red rice, also known as O. sativa f. spontanea, is considered a weedy species in the cultivation of rice, as the reproduction of red rice favours specific environmental conditions (such as flooded fields) that are typical in the cultivation of commercial rice. Outside of rice production areas, red rice is not a weed species. Gene flow from cultivated rice into red rice can occur, although the rate is likely to be very low with levels being dependent on the degree of overlapping of flowering periods. Weedy rice is readily found in tropical America. Weedy rice appears to be mainly composed of annual Oryza spp with feral traits including seed shattering. In contrast to Asia, where manual transplanting is still predominant, direct seeding of weedy rice-contaminated seed is common for a high proportion of rice farmers in tropical America, ensuring field reinfestations and making it one of the most serious weed problems in this region.

Weedy rice is often referred to as red rice because of the red color of its pericarp, and it has been botanically classified as O. sativa f. spontanea, the same species as cultivated rice (Chu et al 1969; Diarra et al 1985; Ellstrand et al 1999; Langevin et al 1990; Oka and Chang 1961). Reports suggest that weedy rice may include other Oryza species, including O. barthii, O. glaberrima, O. longistaminata, O. nivara, O. punctata, O. sativa, andO. latifolia (an American tetraploid) (Holm et al 1997). Hybrid swarms between the American form of O. perennisand O. sativa have been found in Cuba (Chu and Oka 1969). Weedy rice may also have evolved through the dedomestication of cultivated rice to weedy types (Vaughan et al 2003). In addition to seed shattering, weedy rice seeds may possess secondary dormancy, and some types are morphologically indistinguishable from rice varieties yet still shatter seed (Lentini and Espinoza 2005). Natural gene flow estimates in the field from herbicide-resistant rice into weedy rice under temperate conditions indicate hybridisation rates of under one percent (Chen et al 2004; Estorninos et al 2002; Messeguer et 2004; Zhang et al 2003), as confirmed by genetic analysis. However, a cumulative hybridisation rate(over a 3-year period) under temperate conditions may be from 1 to 52% (Guadaggnuolo et al 2001), indicating that genes from rice varieties may transfer and be quickly fixed into weedy rice if they have a selective value. The cumulative rate of introgression may be even higher under tropical conditions, because of the lack of crop rotation and several crop cycles per year. Several biological, genetic, and environmental factors affect the level of outcrossing compatibility, including temperature, humidity, genotype, flower morphology, stigma receptivity, pollen viability, pollen germination, and pollen tube development.

FOOD AND FEED SAFETY CONSIDERATIONS

Antinutrients in rice

Rice contains a small number of antinutritional factors that are concentrated in the bran fraction and which, except for phytic acid, are subject to heat denaturation (inactivation). These antinutrients include: phytic acid, which is a storage form of phosphorus in plant seeds that also chelates calcium, zinc, iron, and magnesium in the digestive tract of animals, thus interfering with the absorption of these nutrients; trypsin inhibitor; and lectins, which are a class of proteins with specific binding affinities for particular carbohydrate moieties of glycoproteins present in cell walls and cell plasma membranes, and which have been associated with a range of antinutritive effects and some disease pathologies.

Phytin:

Phytin is an organic phosphorus compound contained primarily in the bran layer, and it exists as a mixture of calcium-magnesium salts of phytic acid. Free phytic acid (myo-inositol 1,2,3,4,5,6-hexakis dihydrogen phosphate) chelates nutritional metal ions such as calcium and iron, which reduces the absorbability of these ions into the body (Thompson and Weber 1981). It has been reported that phytic acid reduced platelet aggregation and had an inhibitory effect against blood clot formation which may cause thrombosis and atherosclerosis (Vucenik et al 1999). Phytic acid is considered to be an anticarcinogen influencing signal transduction pathways, cell cycle regulatory genes, differentiation genes or suppressor genes (Shamsuddin 1999).

Oryzacystatin:

Oryzacystatin has been isolated from rice bran (Abe et al 1987) and is considered a cysteinyl proteinase inhibitor (cystatin). It is inactivated by heat above 120°C (Juliano 1993).

Lectins:

Lectins are carbohydrate-binding proteins which agglutinate cells that are able to precipitate glycoconjugates or polysaccharides. The toxicity of lectins is due to their ability to bind to specific carbohydrate receptor sites on the intestinal mucosal cells and interfere with the absorption of nutrients across the intestinal wall (Liener 1986). Rice bran lectin, haemagglutinin, has been found to be associated with agglutination of human A, B and O group receptors with specific binding to 2-acetamido-2-deoxy-D-glucose (Poola 1989). Rice bran lectin is heat labile at temperatures above 80°C. Mannnose-binding rice lectin is distributed in all parts of the rice plant, and it has a potential ability to agglutinate bacterial cells of Xanthomonas campestris pv oryzae, the pathogen causing bacterial leaf blight in rice, and also spores and protoplasts of Magnaporthe grisea, the rice blast fungus (Hirano et al 2000).

Allergens

While rice is not considered to be a common cause of allergic reactions to food, allergic reactions have been documented, and certain proteins in rice have been identified as allergens. The first reported allergens in rice were 14-16 kDa proteins which were detected using sera from patients allergic to rice (Matsuda et al 1991). A 16 kDa protein was later recognised as a major rice allergen. This protein has significant amino acid homology to barley trypsin inhibitor and wheat alpha-amylase inhibitor (Izumi et al 1992). Subsequently, rice seed proteins with molecular masses of 26, 33, and 56 kDa have been recognised as being allergenic. The 33 kDa protein has been recently characterised and identified as the enzyme glyoxalase I (Usui et al 2001).

Trypsin Inhibitor:

A trypsin inhibitor has been isolated from rice bran and characterised (Tashiro and Maki 1979). There seems to be no standard way of reporting the quantity of the inhibitor, and it does appear to be heat labile. No trypsin inhibitor was detected in the grain or polished rice, but in the bran (AgrEvo 1999).

Alpha-amylase Subtilisin Inhibitor:

The amino acid sequence of the bifunctional alpha-amylase subtilisin inhibitor from rice is known (Ohtsubo and Richardson 1992). Bifunctional inhibitors have been proposed to be associated with defence of the seed against insect pests and pathogenic microorganisms (Ryan 1990).

GENE FLOW FROM CULTIVATED RICE: ECOLOGICAL CONSEQUENCES

Rice (Oryza sativa) is one of the world's most important cereal crops, providing staple food for nearly one half of the population. In many developing countries, rice is the main source of food security and is intimately associated with local lifestyles and culture. With the rapid increase of global population, much greater rice production is demanded, leading to wide application of transgenic biotechnology to rice for genetic improvement. Although no GM rice has been officially approved yet for extensive commercial cultivation in the world, genes conferring traits, such as high amounts of beta-carotene, high protein content, disease and insect resistance, herbicide resistance, and salt tolerance, have been successfully transferred into different rice varieties through transgenic techniques. Some of these GM rice breeding lines or varieties have been released into the environment for testing. It is apparent that as an important world cereal crop, transgenic rice varieties will inevitably be released into environments for commercial production in the near future.

Undoubtedly, biotechnology and GM crops will provide new opportunities for global food security and development in life sciences. However, the uses of GM crops have also aroused tremendous concerns about their biosafety world wide. The potential ecological risks associated with transgene escape through gene flow are the foremost among these concerns.

When alien transgenes escape to and express in weedy or wild relatives of GM rice, transgenes may persist and disseminate within the weedy or wild populations through sexual reproduction and/or vegetative propagation. Transgenes that are responsible for resistance to biotic and abiotic stresses (such as disease and insect resistance, drought and salt tolerance, and herbicide resistance) can significantly enhance the ecological fitness of weedy and wild populations.

The escape of these transgenes may cause ecological problems, for instance, by producing aggressive weeds, the spread of which might result in unpredictable consequences to local ecosystems. On the other hand, when transgenes escape to and persist in wild rice populations, the rapid dissemination of transgenic hybrid individuals (or progeny) might change the original wild rice populations. In some cases, the aggressive spreading of hybrid swarms with better ecological fitness could even lead to the extinction of endangered wild species populations locally. Therefore, knowledge of the likelihood of gene flow from rice to its weedy and wild relatives will help to predict the magnitude of the potential ecological consequences caused by transgene escape. This knowledge will also facilitate the effective management and safe use of the transgenic crops.

CULTIVATED RICE AND ITS WEEDY AND WILD RELATIVES

The first step in the assessment of gene flow and its consequences is to determine which weedy and wild species can hybridize with the crop to produce fertile offspring. Cultivated rice is included in the genus Oryza of the grass family (Poaceae).

This genus includes two cultivated species (Asian rice Oryza sativa, and African rice O. glaberrima) and more than 20 wild species with ten different genome types, i.e., AA, BB, CC, BBCC, CCDD, EE, FF, GG, JJHH, and JJKK. The wild relatives of rice with different genome types usually have significant reproductive isolation, making them unlikely to hybridize under natural conditions. Therefore, the wild species of concern for transgene escape are only those containing the AA genome.

As close relatives of cultivated rice, some wild rice species such as O. rufipogon, O. nivara, O. longistaminata, and O. glumaepatulaare commonly found or coexist in rice farming systems of many Asian, African, and American countries.

The weedy rice (also referred to as red rice, Oryza spontanea) is frequently observed in rice fields as an accompanying weed, particularly in the rice fields with direct-seeding cultivation practices. These AA-genome weedy and wild relatives are highly compatible sexually with cultivated rice. Their interspecific F1 hybrids could form complete chromosome pairing in meiosis and have relatively high pollen and seed fertility to produce viable offspring. Thus, studying gene flow from rice to its weedy and wild relatives becomes an important component for the potential ecological risk assessment of GM rice, because gene flow is the primary step from which potential ecological consequences of transgene escape may follow.

GENE FLOW FROM CULTIVATED RICE TO ITS WEEDY AND WILD RELATIVES

In order to estimate the pollen-mediated gene flow from cultivated rice to its weedy and wild relatives, experiments were conducted at two sites in Kyongsan of South Korea and Chaling of Hunan Province, China, respectively, under special field conditions mimicking the natural occurrence of weedy and wild relatives in Asia. Two types of experimental designs were established by constructing different populations to examine gene flow from cultivated rice to weedy and wild rice species.

GENE FLOW FROM TRANSGENIC RICE

To weedy rice was measured using a transgenic rice variety (Nam29/TR18, as a pollen donor) with herbicide resistance (bar) and 13 accessions of weedy rice collected from Asia and America. The experimental plot was designed as complete random blocks where Nam29/TR18 was planted and mixed with one of the 13 weedy rice accessions in each block, respectively.

Each block consisted of eight weedy rice plants. For identification of hybrids between Nam29/TR18 and weedy rice, seedlings generated from different weedy rice plants were sprayed with herbicide Basta at the 3—4-leaf stage. The surviving seedlings with resistance to herbicide Basta were considered hybrids and were subject to PCR detection of the herbicide resistance bar gene to confirm their hybridity. Gene flow frequencies were estimated by calculating the number of hybrids against the total number of seedlings germinated.

The average frequencies of weedy rice seedlings with herbicide resistance were very low and varied among different blocks, but with no significant differences among the replications. The experimental results indicated that the detectable rate of herbicide resistance gene flow from the transgenic rice to weedy rice plants varied between 0.011~0.046%.

Gene flow from cultivated rice to perennial common wild rice was measured using the Minghui-63 rice variety, and wild rice O. rufipogon (as pollen recipient) was planted in different models to allow outcrossing to occur naturally.

Co-dominant simple sequence repeats (SSRs) were used as molecular markers for accurate identification of hybrids between cultivated rice and O. rufipogon. The selected SSR primer pair amplified polymorphic alleles from the two species, which were easily distinguishable with electrophoresis in agarose gels.

O. rufipogon presented a consistent fast-migrating allele (F) and Minghui-63 a slow-migrating allele (S) in the gels. The hybrids between the two species displayed stable heterozygous (FS) alleles. Leaf samples of germinated seeds from O. rufipogon populations were collected from individual seedlings for SSR examination.

Gene flow frequencies were estimated by calculating the number of seedlings with the FS heterozygote SSR pattern against the total number of seedlings examined.

As a result, the frequencies of detected interspecific hybrids varied from 1.21~2.94% in different planting models. Gene flow frequency from cultivated rice to O. rufipogon was therefore expectedly high, up to ca. 3%, although humidity and wind strength and direction significantly affected the rate of gene flow.

Fig. 1. Field experiment of gene flow from a herbicide resistant line "Nam29/TR18" of cultivated rice (short plants with yellow panicles) to a Korean weedy rice strain "Galsaegshare" (tall plants with purple panicles) mimicking the natural occurrence of the crop-weed mixture in Asia.

POTENTIAL CONSEQUENCES OF TRANSGENE ESCAPE FROM GM RICE TO ITS WEEDY AND WILD RELATIVES

With current concerns over weed problems caused by wild rice, and particularly the weedy rice in rice farming ecosystems, one of the major fears is whether the transgenes in GM rice varieties will escape to their wild and weedy relatives through gene flow, and enhance the fitness of the wild relatives. This could increase the weediness of wild and weedy rice that invade rice fields, causing serious weed problems.

Our experimental data clearly indicate the likelihood of gene flow from cultivated rice to its wild and weedy species, although with different frequencies. The gene flow frequency from cultivated Minghui-63 to wild O. rufipogon in different planting models varied between 1.1~2.94%. These frequencies are significantly high in terms of transgene escape if the cultivated GM rice varieties are grown in the vicinity of wild rice species. Therefore, for the purpose of preventing or minimizing transgene escape to wild relatives, it is recommended that isolation zones with a sufficient space or with trap plants between GM rice and O. rufipogon should be established, until more effective methods are available. Effective isolation from GM rice will benefit the genetic integrity of in situ conserved wild rice populations.

The detected gene flow frequencies from GM rice line Nam29/TR18 to various weedy rice accessions were very low, ranging from 0.011~0.046% in one generation, when a weedy rice strain occurred simultaneously in a rice field. However, the gene flow frequency from cultivated to weedy rice in large populations might be more significant than the data observed in this experiment. Actually, rice cultivars cross easily with their related weedy forms (red rice) found in direct-seeded paddy fields and produce viable and fertile hybrids with a reasonable rate. In addition, when weedy rice consistently occurs simultaneously with a cultivated rice variety in the same field, the number of hybrids resulting from gene flow could accumulate and increase through generations. If GM rice varieties are released to environments where weedy rice occurs abundantly, the transferred alien genes could spread and accumulate in weedy populations. This may pose a severe problem for weedy rice control and management in rice production. Therefore, release of transgenic rice with genes that can significantly increase weediness and can resist weed control measures is not recommended in regions where weedy rice is already a serious weed problem.

WEEDY RICE

Weedy rice, also known as red rice, is a variety of rice (Oryza) that produces far fewer grains per plant than cultivated rice and is therefore considered a pest. The name "weedy rice" is used for all types and variations of rice which show some characteristic features of cultivated rice and grow as weeds in commercial rice fields. Populations of weedy rice are found in many rice-growing regions. Weedy rice varieties generally have fragile stalks that self-seed before harvest. Variations of weedy rice adapt to a wide range of natural conditions.

Weedy rice grains often have a red pericarp, so for this reason in the international literature the term "red rice" is often used. This term, however, is not very relevant, because the rice with a red pericarp are also found in some cultivated varieties, and is absent in many forms of weedy rice.

In most regions of the rice production, weedy rice is introduced after the transition from transplanting rice to sowing commercial seeds directly in the rice field. It has become very significant since the mid-1980s, especially in Europe, with the weak semi-dwarf varieties of indica subspecies. Distribution generally promotes use of commercially purchased seeds containing weed seeds. Weedy rice can be found in 40-75% of rice fields in Europe, 40% of fields in Brazil, 55% in Senegal, 80% in Cuba, and 60% inCosta Rica.

Because weedy rice and cultivated rice are so closely related, herbicides that would kill red rice would also kill cultivated rice. A genetically modified form of cultivated rice has been developed that will resist a herbicide, but this form of rice has not been approved for human consumption. This genetically modified form of cultivated rice has, however, appeared on the rice market. One way sometimes used of combatting weedy rice is to plant a red rice one year and on that land a green rice next year, and so alternating, and in each crop to pull up any rice plants that are the wrong color for that year.

PHENOTYPIC SELECTION FOR DORMANCY INTRODUCED A SET OF ADAPTIVE HAPLOTYPES FROM WEEDY INTO CULTIVATED RICE

Association of seed dormancy with shattering, awn, and black hull and red pericarp colors enhances survival of wild and weedy species, but challenges the use of dormancy genes in breeding varieties resistant to preharvest sprouting. A phenotypic selection and recurrent backcrossing technique was used to introduce dormancy genes from a wild-like weedy rice to a breeding line to determine their effects and linkage with the other traits. Five generations of phenotypic selection alone for low germination extremes simultaneously retained dormancy alleles at five independent QTL, including qSD12 (R2 > 50%), as determined by genome-wide scanning for their main and/or epistatic effects in two BC4F2populations. Four dormancy loci with moderate to small effects colocated with QTL/genes for one to three of the associated traits. Multilocus response to the selection suggests that these dormancy genes are cumulative in effect, as well as networked by epistases, and that the network may have played a "sheltering" role in maintaining intact adaptive haplotypes during the evolution of weeds. Tight linkage may prevent the dormancy genes from being used in breeding programs. The major effect of qSD12 makes it an ideal target for map-based cloning and the best candidate for imparting resistance to preharvest sprouting.

WEEDS, plants adapted to human disturbances, compete with crops for environmental resources and occasionally interbreed with conspecific cultivars. Understanding the genetic basis underlying adaptation is crucial to devising integrated strategies for weed control or to preventing the rise of "superweeds," that is, weeds with transgenes derived from transgenic crops. Conversely, to meet breeding objectives for sustainable agriculture, conspecific weeds, as part of the primary gene pool, are valuable to regain some genetic diversity that was eliminated during domestication. Weedy rice, including red rice, accompanies cultivated rice worldwide. Research on weedy rice has been done for its origin, classification, control, and risk assessment related to transgenic cultivars, but rarely on the genetics of adaptive traits or for identification of beneficial genes for rice improvement.

Seed dormancy optimizes timing of germination for wild and weedy plants and provides resistance to preharvest sprouting (PHS) for cereal crops. Both dormancy and PHS are complex traits controlled by many genes or quantitative trait loci (QTL) such as in Arabidopsis, barley, rice, sorghum, and wheat. Dormancy alleles at a few QTL have been introduced into the nondormant genetic background to validate their effects or to explore their potential in breeding programs. Validated QTL may be cloned to characterize molecular mechanisms directly regulating germination and dormancy.

Utilization of dormancy genes from wild and weedy germplasm to control PHS in cereal crops may be hampered by linkage with some traits that may have an adaptive value under natural conditions but are undesirable for modern cultivars. Dormancy association with red grain color in wheat has prevented the use of these dormancy genes in the development of white grain-colored cultivars with resistance to PHS. Dormancy is also associated with seed appendages, or shattering, and black pigmentation in other grass species. Some of the associations in wild (Oryza rufipogan) and weedy (O. sativa) rice are explained by QTL clustered on the same chromosomal blocks. Introduction of the QTL regions into breeding lines, i.e., nondormant pure lines, enables a precise assessment of the linkage strengths, and it is the first step in characterizing the structure of dormancy gene-related linkage disequilibrium and in understanding the evolutionary history of the adaptive haplotypes across grass genomes.

Phenotypic selection with recurrent backcrossing was proposed to isolate genes with a major effect on a quantitative trait and is now combined with QTL analysis to simultaneously discover and transfer useful alleles from nondomesticated germplasm into breeding lines or intermediate breeding materials.

Genetic analysis suggested the presence of major dormancy genes in some weedy rice accessions. Thus, we used a phenotypic selection technique, without assistance with molecular markers or morphological characteristics, to initiate introduction of dormancy genes from a weedy accession into a breeding line. After completion of dormancy QTL mapping (GU et al. 2004), the linkage map was employed to determine QTL retained by the phenotypic selection over five generations. Five dormancy QTL regions, including four haplotypes for other adaptive traits, were introduced into the background of a breeding line.

MATERIALS AND METHODS

Plant genotypes and mating scheme:

The breeding scheme was used to introduce dormancy genes from the weedy accession SS18-2 to the EM93-1 background. A rice seed usually refers to a dispersal unit that consists of the maternal (i.e., hull, pericarp, and testa) and offspring (i.e., endosperm and embryo) tissues (GRIST 1986). SS18-2 is a wild-like indica-type weedy rice originally collected from Thailand and has a high degree of seed dormancy and shattering, long awns, dark pigmentation on the hull, and red pigmentation on the pericarp/testa. EM93-1 and CO39 are indica-type early and moderate maturation lines, respectively, and do not have dormancy, shattering, awn, and black or red pigmentations on the maternal tissues. Selection started with the F2 CO39/SS18-2, the population with the highest estimate of heritability for dormancy in the experiment (GU et al. 2003).

A strongly dormant, early flowering F2 segregate (i.e., plant 14, ~30 days earlier than the earlier flowering parent CO39) was selected to cross with EM93-1. Individuals having the lowest germination level in the F2-derived F1 to BC3F1 populations were backcrossed with EM93-1 to develop the next generation. EM93-1 was used to replace CO39 as the recurrent parent in the backcross because this breeding line has the duration to flowering similar to that of the above F2 plant 14 under long day lengths and was used to develop the population to map the dormancy QTL. Hybridizations were made using ratooning plants from the selected individuals in each generation.

Similar to SS18-2, the low germination extremes selected in early generations also displayed other weedy traits, although awn length shortened, darkness of the hull lightened, and degree of shattering decreased with each generation. In the BC4F1 generation, plants 44 and 132, which had the traits identical to EM93-1, other than those mentioned above, were selected to develop F2 populations, i.e., BC4F2 (44) and BC4F2 (132). Plant 132 seeds had the lowest germination rate and red pericarp color, and plant 44 seeds had a moderate level of germination and shattering, relatively long awns, black hull color, and red pericarp color. These BC4F2's were used to determine the effects of retained dormancy QTL and linkage with other weedy characteristics.

Plant cultivation and phenotypic identification:

The populations were grown in the greenhouse from 2000 to 2004. Plants were cultivated in pots (28-cm diameter × 25-cm height), with one plant per pot filled with a mixture of clay soil and SUNSHINE medium (Sun Gro Horticulture Canada, Seba Beach, AB, Canada).

Day/night temperatures were set at 29°/21° and the day length was set for 14 hr. Seeds were harvested at 40 days after flowering, which was measured by emergence of the first panicle in the plant. Seeds were cleaned and air dried in the greenhouse for 3 days to ~12% moisture content. Dried seeds were sealed in plastic containers and stored at -20° to prevent them from after-ripening prior to use.

The degree of dormancy was measured by percentage of germination. Prior to germination, seeds sampled from each plant were after-ripened at room temperature (23°–25°) for different periods of time (1–45 days). Three replications of ~50 seeds each were placed in 9-cm petri dishes that were lined with a Whatman no. 1 filter paper, wetted with 10 ml deionized water, and incubated at 30° and 100% relative humidity in the dark for 7 days. Germination was evaluated visually by =3-mm protrusion of the radicle or coleoptile from the hull. Percentage of germination (x) was transformed by sin-1(x)-0.5 for statistical analysis.

Phenotypes for awn, black hull color, and red pericarp color were identified on the basis of the presence or absence of each characteristic from the F2 to BC4F1 generations, as in previous research (GU et al. 2003); the presence and absence were scored as 1 and 0, respectively, for correlation analysis. There was no difficulty in distinguishing the red pericarp color genotypes from the white ones on the basis of the appearance of fully matured seeds; thus this trait was also scored as the presence and absence in the two BC4F2's. The other weedy traits in the BC4F2 (44) population were quantified by awn length, intensity of component pigmentations, and shattering rate for QTL analysis (GU et al. 2005b).

Briefly, the awn length was expressed as the mean length averaged over three samples of 50 seeds each. Shattering rate was expressed as the percentage of shattered to total air-dried seed weight. To assess shattering, panicles were cut from the plant and gently shaken for ~20 sec over a container to collect shattered seeds and then hand threshed to collect nonshattered ones. Pigmentation was expressed as spectral reflectance. Reflectance was measured with a Chroma Meter (Minolta CR310). The Chroma Meter decomposes reflectance spectra of hull color into three (i.e., L, a, and b) dimensions, with low L and high a or b positive values indicating a high intensity of black and red or yellow pigmentations, respectively.

QTL analysis:

The selected plants from the F2 to BC3F1 generations were genotyped with rice microsatellite (RM) markers flanking the six dormancy QTL to track the SS18-2-derived alleles. Genomic DNA was prepared from ~50 seedlings bulked from individual F2 to BC3F1 plants. BC4F1 plants 44 and 132 were genotyped for 140 RM markers evenly distributed over the framework linkage map to scan for chromosome (chr) segments from SS18-2. Genomic DNA for the BC4F1 plants was prepared from the leaves. Additional markers were screened and were used to delimit the chr segments. Two populations of BC4F2 plants were genotyped for all the polymorphic markers retained in the BC4F1 plants. Genomic DNA from individual BC4F2 plants was prepared from young leaves. DNA was extracted, the markers were amplified by polymerase chain reaction (PCR), and the products were resolved using methods previously described. Markers were positioned using MAPMAKER/EXP 3.0.

One- or two-way ANOVAs were used to detect QTL segregating in BC4F2 populations. These analyses were based on a linear model in which a phenotypic value was partitioned into the mean, genotypic, and error (including random error and the residual effect unexplained by the genotypic effects) components.

The one-way ANOVA was performed for all markers retained on each SS18-2-derived segment. The marker that contributed most to the phenotypic variance as compared with the other markers on the same segment was used to estimate its epistasis with the marker on another SS18-2-derived segment. For two-way ANOVA, the genotypic effect in the model for the above one-way ANOVA was further partitioned into main and interaction effects of individual loci. The threshold for a significant main or epistatic effect was set at the 5% probability level. The contributions (R2) of the main or epistatic effects were calculated as the proportion of component type III sum-of-squares (SS) to the corrected total SS. The software WinQTLCart was used to infer the relative order for QTL located in the same chr region.

RESULTS

Responses to phenotypic selection:

Individuals selected from the F2 to BC3F1 populations were dormant genotypes because segregation for germination (0–100%) occurred in the next generation. Segregation also occurred for the traits awn, black hull color, and red pericarp color in all of the above generations. The association between dormancy and red pericarp color was not significant in the F2 plant 14-derived F1 population, which was similar to that in the F2 CO39/SS18-2 population, but was significant in the BC1F1 to BC4F1 generations. There is a dormancy QTL (i.e., qSD7-1) allele linked in coupling with the red pericarp color gene Rc. Increase in strength of the association with advance of the BC generations, or with replacement of the CO39 with the EM93-1 genome as a result of backcrossing, confirms the linkage and suggests that the parent CO39 must contain factor(s) elsewhere in the genome affecting expression of this dormancy gene.

Phenotypic correlations of dormancy with awn and black hull color also differed by generation. The differences might arise from variations in expressivity (i.e., percentage of seeds with an awn or awn length and hull color darkness) of these morphological traits in the populations.

In addition to the qSD7-1 region, selection in early generations also retained dormancy alleles at qSD4, -8, and -12 as indicated by the presence of

SS18-2-type alleles at markers flanking the three QTL. Plant 14 from the F2 CO39/SS18-2 population was homozygous for CO39-type alleles at two qSD6 flanking markers, indicating that the initial selection lost this dormancy allele. Plants selected from the F2 to the BC2F1generation had the SS18-2-type allele at only one of two markers flanking qSD7-2; we surmise that this dormancy allele could be lost either in the initial selection or in selections before the BC4F1 generation.

The genome-wide scan detected five SS18-2-derived chr segments for BC4F1 plants 44 and 132. These plants differed completely in segments on chr 4, 8, and 12, had overlapping segments on chr 7 and 10, and shared a segment on chr 1 as defined by the RM markers in the regions. The length of donor chr segments varying from ~10 to 40 cM and the five segments accounted for ~6.6% of the SS18-2 haploid genome. The proportion of donor segments is close to the expectation (6.25%) for a BC4 generation. Previous research did not detect dormancy QTL on the chr 1 and 10 segments; their presence in the BC4F1's suggests they may also harbor dormancy genes or associate with some unidentified factors related to the selection regime.

Dormancy QTL:

The BC4F2 populations differed in distribution patterns for germination. For example, it took the BC4F2 (132) and BC4F2 (44) populations ~25 and 20 days of after-ripening (DAR), respectively, to display the largest variation in germination and >45 and ~30 DAR, respectively to reach 80% mean germination. In addition, the BC4F2(132) and BC4F2 (44) populations displayed bimodal and nearly normal distributions, respectively, when the largest variation occurred. These results suggest that the dormancy alleles differentiated between BC4F1 plants 44 and 132 may also differ in the magnitude of their gene effects.

Dormancy QTL were detected on the chr 1, 7, and 12 segments by their main or epistatic effects from BC4F2 (132). The QTL on chr 7 and 12 coincide with the previously identifiedqSD7-1 and qSD12, respectively. The locus qSD12 accounted for ~50% of the phenotypic variance and its genetic effect was predominantly additive. The locus qSD7-1 accounted for ~9% of the phenotypic variance and consisted of both gene additive and dominance effects. The locus qSD7-1 also was involved in digenic epistases with qSD12 and a locus near marker RM220 on the chr 1 segment; the component two-way interaction effects accounted for 4.6% ($P = 0.020$, data not shown) and 6.4%, respectively, of the phenotypic variance. The locus near RM220 was not detected in the primary segregating population and is now designated qSD1.

Dormancy QTL were detected on the chr 1, 4, 7, and 8 segments by their main or epistatic effects from BC4F2 (44). The QTL on chr 4, 7, and 8 coincide with the previously identifiedqSD4, -7-1, and -8, respectively. The locus qSD1 in this population had a relatively small main effect (R2 ˜ 7%). It was also involved in digenic epistases with both qSD7-1 and qSD8, where the component two-way interactions accounted for 4.7% (P = 0.047) and 9.1% of the phenotypic variance, respectively. Intriguingly, the epistases involving qSD1 displayed different patterns. The difference in degree of dormancy between two homozygous genotypes at qSD1 was greater when the dormancy alleles were present at qSD7-1 or absent at qSD8.

Haplotypes:

The loci qSD7-1 and Rc colocated in the region between markers RM5672 and RM180 on the basis of the two BC4F2 populations. The Rc locus, which colocates with markers E10534S or aligns with RM5672 in high-resolution maps, was estimated to be 1.5 cM distal from RM5672 on the basis of 390 individuals from the BC4F2 populations. The dominant gene Rc contributed a few percentiles more to phenotypic variance in germination than the marker RM5672 in both BC4F2 populations. The reason that RM5672 as the marker nearest qSD7-1 is because its codominant nature facilitated estimation of gene additive and dominant effects for the locus.

A range of variation in awn length, shattering rate, and intensity of component pigmentations on the hull occurred in the BC4F2 (44) population. The distribution patterns for these traits (data not shown), especially the awn length and pigmentations, were similar to those in the primary segregation population. Two QTL for awn length were detected on the chr 4 and 8 segments, respectively, and their main and epistatic (17.4%, P < 0.0001) effects together accounted for 78.8% of the phenotypic variance. The same segments also harbored QTL for shattering. Two QTL for hull color were detected on the chr 1 and 4 segments, respectively; at both loci the SS18-2- and EM93-1-type alleles increased the intensities of black and yellow/red pigmentations, respectively. The QTL for awn, shattering, and hull color on chr 4 and 8 were also detected in the primary segregation population, but they link more tightly with dormancy QTL in the BC4F2 (44) population as indicated by common nearest markers. The locus for hull color on chr 1 is designated as qHC1, as it was not detected in the primary segregating population. The BC4F2 (132) population also segregated for the qHC1 region, but not for hull color as judged by visual examination; we measured intensities of the three-component pigmentations for 140 BC4F2(132) plants using the same method, but failed to detect a significant effect of qHC1 (R2 < 0.01, P > 0.9) in the subpopulation. Considering its relatively minor effect on black pigmentation (R2 < 5%), we conclude that qHC1 is a modifier to the major locus qHC4 (R2= 50%).

The quantification methods used here improved the correlations of degree of dormancy to the awn, shattering, and hull color traits in the BC4F2 (44) population, as compared with the results from early generations. The above advanced BC analysis demonstrated that the correlations arose from the tight linkage among/between QTL on the chr 1, 4, 7, and 8 segments. The four sets of the SS18-2-derived haplotypes most likely locate within the genetic distance of a few centimorgans because the QTL for different traits on each of the segments shared common nearest markers and identical peak positions as inferred by the WinQTLCart mapping (data not shown).

DISCUSSION

Dormancy gene action over generations of selection:

At least five dormancy QTL, including previously identified qSD4, -7-1, -8, and -12, responded to the single-plant selection in the early generations. Additive effects varied from insignificant (e.g., qSD8) to a magnitude equivalent to 1 unit of standard deviation (i.e.,qSD12) in the relatively synchronized genetic backgrounds. This multilocus response was much higher than the theoretical probability (1/32 or 3.1%) for a five-locus heterozygote in a BCF1 population and was even higher than the expectation for segregating loci retained by the selection with recurrent backcrossing technique. We selected for low germination extremes and determined that most QTL consisted of predominately gene additive effects.

Thus, we attribute the simultaneous retention of the five dormancy alleles for the most part to the gene cumulative effect postulated by CHANG and TAGUMPAY (1973) for dormancy in rice. In addition, the two- or higher-order interactions in the primary (GU et al. 2004) and the BC4F2 segregation populations suggest that all dormancy loci identified from SS18-2 are networked by epistases.

It is likely that the phenotypic selection was more favorable for networked genes rather than for a set of random loci. For example, the qSD1 dormancy allele did not display a significant main effect in the BC4F2 (132), yet it was retained during selection. Thus, it must have benefited from its epistasis with qSD7-1, a moderate locus (R2 = 9%) that also interacted with the major QTL qSD12. Similarly, the dormancy allele at qSD8 retained in BC4F1 plant 44 benefited from its epistasis with qSD1, which also interacted with the qSD7-1. Further research will be needed to isolate QTL, which have minor effects or only epistatic effects, from multilocus systems as single Mendelian factors to determine their genetic nature.

Implications of dormancy gene-related haplotypes:

Four haplotypes define the genetic basis for associations of seed dormancy with shattering, awn, black hull color, and red pericarp color traits over generations. Although dormancy alleles in the haplotypes had moderate to small effects, their transmission directly affected the fate of groups of alleles, including major genes for other weedy traits in a population. The adaptive significance of dormancy in seed-bearing plants has been limited to the promotion of survival under adverse environmental conditions by distributing germination timing. The other weedy traits also contribute to adaptation in different ways, such as shattering enables weed seeds to escape from harvest, long awn aids seed dispersal, and chemicals underlying hull and pericarp colors aid in seed persistence in the soil. In adverse environments, genotypes with a high level of seed dormancy due to having multiple dormancy genes with epistatic effects can survive longer; these strongly dormant genotypes are those carrying genes for the other weedy traits because of tight linkage in the haplotypes. Our phenotypic selection, as a simulation of natural selection for strongly dormant genotypes, suggests that the multiple-gene system governing dormancy sheltered genes for other important adaptive traits during the evolution of weeds.

Haplotypes are signatures for evolutionary genetics or comparative genomics. Wild-like weedy rice such as SS18-2 is considered to originate from a natural hybridization between wild Oryza ssp. and cultivars on the basis of morpho-physiological characteristics. QTL or QTL clusters for some of the traits we measured have been reported in O. rufipogenaccessions, suggesting that the haplotypes detected in SS18-2 were likely inherited as units from their wild relative, rather than by the hitchhiking effect of allelic mutations after differentiation of the wild and weedy species.

The origin of cultivar-like weedy rice, such as red rice, which is distributed in East Asia, America, and Europe where there was no wild rice (O. ssp.), remains uncertain. Examining the haplotypes in cultivar-like weedy rice should be helpful in determining its origin. In addition, the weedy traits studied in this research are also common in other grass species (HARLAN et al. 1973). Some genes for shattering and red pigmentation have been used to develop a consensus map for grasses. Isolation and characterization of these adaptive haplotypes should enhance our knowledge about evolution of grass genomes and weediness.

There are increasing concerns about the risks of gene flow from transgenic cultivars to conspecific weeds. Natural hybridization occurs between weeds and cultivars and initiates hybridization-differentiation cycles (LADIZINSKY 1985; OKA 1988). Once transgenes enter the cycles, the transgenic recombinants have greater opportunity to survive because of seed dormancy. Furthermore, if transgenes integrate into a weedy haplotype region, the recombinants could become superweeds. GRESSEL (1999) proposed that construction of a transgene (e.g., herbicide-resistant genes) in tandem with a gene for nondormancy or nonshattering would mitigate the risk of transgenic cultivars becoming "volunteer" weeds in the following crop. Theoretically this proposal has merit for managing the incidence of transgenic weeds, but technically many issues must be resolved. For example, there is no information on the molecular structure of genes that directly regulate dormancy because they have not been cloned; there is insufficient information on what dormancy locus region might be the best target for insertion of a transgene as it relates to the effects of genes in natural populations and the effects of genetic background due to epistasis; and, most importantly, there is no information about how the genes for weedy traits evolved to form haplotypes as it relates to the possibility of developing superweeds with an additional trait such as herbicide resistance.

Challenges and promises for the use of dormancy genes:

Domestication and breeding activities have eliminated a substantial degree of seed dormancy from modern cultivars by selection for rapid and uniform germination; thus, PHS has been a worldwide problem in agriculture. Linkage drag with undesirable traits like shattering, awn, black hull color, and red pericarp color is a major challenge for the use of weedy rice-derived dormancy genes for improving resistance to PHS. It will be difficult to separate some dormancy alleles from the linked genes as indicated by strong associations over successive generations.

Association between red pericarp color and seed dormancy in cereal grains could be due to pleiotrophy or tight linkage of genes. Fine mapping of theqSD7-1 region or cloning of the dormancy locus will reveal the nature of the association and, therefore, determine if this dormancy allele could be used in breeding. The same approach would also be necessary to determine the usefulness of dormancy genes in other haplotypes. Additional challenges include the variation in gene effects with genetic background and the genotype-by-environmental interaction. For example, qSD1 was not detectable in the primary segregating population, but was detected in the BC4F2 (44) population. It is not unusual that a dormancy allele could offset or even reverse the effect of another dormancy allele on germination due to epistasis.

Major dormancy QTL, such as qSD12, are promising candidates for breeding varieties resistant to PHS and convey key genetic information on the regulation of germination and after-ripening. The major effect of qSD12 was detected in the primary segregation population grown in different years and it had no deleterious effects in the BC4F2 (132) population. The large additive effect suggests that gene(s) underlying qSD12 can be incorporated into conventional varieties or parental lines of hybrid rice to markedly improve resistance to PHS.

QTL that explained =50% of the phenotypic variance in dormancy or dormancy-related traits (e.g., amylase content) have been identified from two- and six-rowed barley varieties. Fine or comparative mapping suggests that barley QTL may consist of a gene cluster and may be conserved in rice and wheat. We are developing isogenic lines for qSD12 to clone its underlying gene(s) by taking advantage of the published rice genome sequences.

GENETIC DIVERSITY AND POPULATION STRUCTURE OF WILD/ WEEDY EGGPLANT (SOLANUM INSANUM, SOLANACEAE)

Wild and weedy relatives of crops constitute a key source of genetic diversity that can be tapped by plant breeders to obtain novel traits for crop improvement (Harlan, 1976). For example, many previous studies have exploited populations of crop relatives to obtain genes that confer traits such as resistance to pathogens and pests and tolerance of harsh abiotic conditions. Research on the abundance and distribution of crop relatives has been the focus of many efforts to enhance the germplasm available for the world's food supply (e.g., Maxted et al., 2012).

However, despite this well-recognized value, most wild and/or weedy relatives of crops remain largely understudied and their natural populations are increasingly at risk from multiple threats, including climate change, habitat loss, "genetic swamping", and displacement by invasive plants. Estimates suggest that only 2–4% of the world's ex situ germplasm collection of major food crop gene pools comprise wild/weedy relatives (Rao, 2013). Knowledge of the nature and pattern of genetic variation in extant populations of wild/ weedy relatives can aid in designing effective strategies for their conservation. For example, such studies can be used to identify regions with genetically distinct populations, those that are genetically depauperate vs. extremely diverse, and those that have hybridized with other taxa, including related crop varieties (Nevo, 1998; Rao and Hodgkin, 2002; Guo et al., 2012). In general, crop wild relatives represent a unique component of natural biodiversity, as well as interesting examples of how population structure can be influenced by human activities.

In this study, we focus on wild/weedy relatives of cultivated eggplant, S. melongena L., known as "brinjal" in India, one of its centers of diversity. A global analysis of the conservation status of 81 important crop gene pools ranked eggplant second among crops whose wild gene pools are greatly underrepresented in ex situ holdings, and therefore warrant urgent conservation efforts (http://www.cwrdiversity.org/conservation-gaps/; also seeVincent et al., 2013). Eggplant is grown worldwide for its edible fruits and is one of the most important vegetables in Asian diets. It is widely agreed that eggplant originated in tropical Asia, but particular centers of domestication are still debated, as is its closest wild progenitor. Both India and China have been suggested as the most likely centers of eggplant domestication (Lester and Hasan, 1991; Daunay et al., 2001a, b; Doganlar et al., 2002; Bhat and Vasnathi, 2008; Meyer et al., 2012).

Questions about the domestication of eggplant have been complicated by a long-running lack of clarity about the taxonomy of the crop and its most closely related wild/weedy species. Briefly, many authors have followed the taxonomic approach of Lester and Hasan (1991) and grouped wild eggplant as a variety of the crop (S. melongena var. insanum; e.g., Mace et al., 1999; Behera et al., 2006; Weese and Bohs, 2010; Tümbilen et al., 2011; Daunay and Hazra, 2012; Samuels, 2012, 2013). Alternatively, Knapp et al. (2013) recently classified cultivated eggplant and its closest wild relatives into 10 formal species. They unified the S. melongena groups E (wild) and F (weedy) of Lester and Hasan (1991) into a single species under the name S. insanum L., as inKarihaloo and Rai (1995) and noted that that the crop and S. insanum can be considered as separate taxa because they have different selection regimes.Knapp et al. (2013) stated that S. insanum "is a wild plant, although often growing in disturbed areas including those around villages". They postulated that S. insanum is "almost certainly" the wild progenitor of the domesticated eggplant, with which it is completely interfertile and that it may include feral reversions of domesticates. Recognizing valid reasons for both classification systems (either S. insanum or S. melongena var. insanum), we use S. insanumbelow. Furthermore, we have not attempted to divide the "wild" populations into discrete categories such as wild, weedy, crop–wild hybrids, or crop-derived feral populations, which are not easy to distinguish. Below, we refer to the crop, S. melongena, as "brinjal" and to S. insanumas "wild brinjal".

India is known to harbor a wide diversity of local landraces of brinjal and its wild relatives (Deb, 1989; Mace et al., 1999; Karihaloo et al., 2002). Early genetic analyses of Indian materials using allozymes and RAPD markers reported higher levels of genetic diversity within populations of wild brinjal than the cultivated S. melongena. These authors also found close genetic similarity between cultivars and "S. insanum", which led Karihaloo et al. (1995) to question the validity of a taxonomic subdivision between them. Other previous genetic analysis studies were either based solely on samples of cultivated S. melongena (Prohens et al., 2005; Munoz-Falcon et al., 2009; Demir et al., 2010; Verma et al., 2012;Cericola et al., 2013; Ge et al., 2013) or incorporated only a few samples of the closely related wild and/or weedy types (Karihaloo et al., 1995; Mace et al., 1999; Nunome et al., 2003; Furini and Wunder, 2004; Behera et al., 2006;Stagel et al., 2008). The studies to date were conducted mostly with historical ex situ accessions whose patterns of genetic variation may not accurately reflect those in extant natural populations. Furthermore, previous studies mainly examined one or a few plants per accession. Thus, the previous studies did not adequately examine wild brinjal genetic diversity at the within- or between-population level to describe its magnitude or its spatial structure.

Mechanisms of pollen- and seed-mediated gene flow are expected to influence the partitioning of genetic diversity within and among wild populations (e.g.,Hamrick and Godt, 1996). Wild brinjal is a widespread, self-compatible perennial that relies on insect pollinators for seed set (Daunay and Hazra, 2012; Davidar et al., 2015). The plants have both functionally male and hermaphroditic flowers, with the latter having stigmas that extend beyond the anthers, a trait that likely promotes outcrossing. We have observed bees such as Amegilla sp., Xylocopa sp., Nomia sp., and Heterotrigona iridipennisforaging for pollen on flowers of both wild and cultivated brinjal. Wild brinjal produces fleshy, bitter, multiseeded, round fruits, that are sometimes used for food and Ayurvedic medicinal treatments (e.g., Kudlu and Stone, 2013). The seeds are probably dispersed by domestic livestock and wild vertebrates, as well as people.

Here, we studied 10 wild brinjal populations across ~600 km in southern India to gain a better understanding of the underlying ecological and evolutionary processes that shape patterns of genetic diversity in wild germplasm. The specific goals of our study were to use polymorphic SSR markers to examine population structure, outcrossing rates, and genetic diversity within and among extant populations. We also sampled three local crop populations in close proximity to their wild/weedy counterparts to investigate the extent of genetic divergence and gene flow between these closely related taxa.

MATERIALS AND METHODS

Study populations and plant materials

We studied wild brinjal populations in the Western Ghats, a mountainous escarpment that runs roughly parallel to the coast of southwestern India. In this region, wild brinjal occurs as a spiny, perennial shrub that is sometimes decumbent and often grazed by vertebrates, typically along roadsides, in wastelands, and in sparsely vegetated areas near villages and agricultural fields. The native vegetation near these populations is mainly a thorny scrub, and the climate is characterized by an 8–9-mo dry season.

We sampled 10 natural populations of wild brinjal across two states in southern India, Karnataka and Tamil Nadu. Elevations of the study populations ranged from 97 to 925 m a.s.l.. In November 2011 and January 2012, we collected wild brinjal DNA samples from sites that ranged in proximity from ~4.5 km apart (sites 8 vs. 9) to ~634 km apart (sites 1 vs. 10). Relative population size was estimated by counting the wild brinjal plants found within ~1 km of the area where DNA was sampled. In each population, we sampled fresh leaves from 20 to 24 individuals that were located at least 5 m apart. Photos were taken of each population, and voucher specimens collected from sites 5, 9, and 10 were deposited at the herbarium of the National Bureau of Plant Genetic Resources in New Delhi (specimen numbers 21552–21554).

We also sampled cultivated brinjal near wild populations at sites 1–3. Cultivated plants at sites 1 and 2 were grown on small farms. At site 1, cultivated brinjal was locally known as Manjiri Gota, with medium-sized, oval fruits that were purple with white stripes when harvested. Fruits at site 2 were medium-sized, oval or round, and purple with green and white stripes when harvested. Cultivated plants at site 3 were sampled from a kitchen garden and had small, round fruits. We do not know the local names for brinjal sampled at site 2 or 3.

DNA was collected from young leaves using 3 × 10 cm cards made from Whatman Grade 3 mm chromatography paper (GE Heathcare, Pittsburgh, Pennsylvania, USA) as done by Adugna et al. (2011). A fresh apical leaf was excised from the plant and squashed onto the card (folded in half to sandwich the leaf in between) using a hammer. To avoid cross-sample contamination, we used fresh polythene paper to cover the cards each time the squash was made. Subsequently, the cards were stored in separate plastic zip-lock bags containing silica gel to facilitate drying. Permits for exporting DNA samples to the United States for simple-sequence repeat (SSR) analysis were obtained from the National Biodiversity Authority of India, as well as the appropriate state-level authorities. We could not collect samples from a larger number of plants, populations, or regions due to limitations imposed by the lengthy and unpredictable permitting process. The SSR data used in this study are recorded in the Dryad Digital Repository: http://doi.org/10.5061/dryad.j5bn2.

DNA extraction, PCR analysis, and genotyping

Genomic DNA was isolated from the Whatman chromatography cards following the method described by Adugna et al. (2011). PCR was performed using 14 SSR primers in multiplexed reactions. The forward primers were labeled with NED, HEX, or 6-FAM fluorescent dyes (Applied Biosystems, Foster City, CA, USA). PCR was performed in a final reaction volume of 10 μL, containing 6 μL Qiagen multiplex PCR master mix, 1 μL ddH2O, 2 μL of primer mix (each primer at 1 μmol/L), and 1 μL DNA template. The PCR program consisted of 95°C for 15 min, followed by 11 touchdown cycles at 95°C for 30 s, 66°C (-1°C/cycle) for 90 s, and 72°C for 60 s. Those cycles were followed by 30 cycles at 95°C for 30 s, 58°C for 90 s, 72°C for 60 s; and a final extension step of 30 min at 72°C. PCR products were diluted 1:20 in ddH2O. For allele sizing, 1 μL of diluted product was combined with 0.5 μL of a universal Rox-labeled internal size standard and 15 μL HiDi Formamide (Applied Biosystems), and separated on an ABI Prism 3100 genetic analyzer (Applied Biosystems). Peaks were detected and alleles scored using the software GeneMapper 3.7 (Applied Biosystems).

Data analysis

For population genetic parameters, we estimated the total number of alleles (AT), observed heterozygosity (Ho), unbiased expected heterozygosity (He), and fixation index (FIS) using the program GENETIX. Outcrossing rates (t) of the populations were estimated from the fixation index estimates using the equation t = (1 - FIS)/(1 + FIS) (Weir, 1996). We also quantified allelic richness (Ar) and private allelic richness (Ap) using the program HP-Rare (Kalinowski, 2005). Differences among populations in basic population parameters were evaluated for significance using Kruskal–Wallis tests in the software R.

Genetic differentiation among populations was assessed by computing pairwise and global FST (Wright, 1949) using the unbiased method of Weir and Cockerham (1984) as implemented in GENETIX. To investigate how genetic diversity was partitioned among the 10 wild brinjal populations, analysis of molecular variance (AMOVA) was performed in GenAlEx version 6. We used the method of Rousset (1997) to test for isolation by distance among the wild populations.

First, the geographic distance matrix was generated using the software GenAlEx. A matrix of pairwise genetic distances [FST/(1 - FST)] among the populations was computed and regressed against a matrix of pairwise logarithmic distances among the sample sites using the software SPAGEDI.

Significance of the regression slope was tested by 1000 random permutations of sampled sites under the null assumption of no correlation between genetic distance and geographic distance.

The program STRUCTURE version 2.3 was used to infer the number of genetic groups (K) in the data set, and to assign individuals among the groups, based on Bayesian analysis.

We used an "admixture" model, with correlated allele frequencies and no a priori sample information, to estimate the number of genetic groups ranging from K = 1 to K= 10. Each STRUCTURE run was performed with 100000 burn-in Markov chain Monte Carlo (MCMC) iterations, and 100000 data collection MCMC iterations, with 5 runs per K value.

Subsequently, the method developed by Evanno et al. (2005) and implemented in the program STRUCTURE HARVESTER version 6.0 (Earl and vonHoldt, 2012) was used to detect the "most likely" number of genetic clusters (K).

To visualize patterns of association among wild and crop populations of brinjal, we performed principle component analysis (PCA) using the ADEGENET (Jombart, 2008) and ADE4 packages of the software R. Hierarchical cluster analysis was carried out in DARWIN 5.0 software, based on a pairwise Nei's genetic distance (Nei, 1972) matrix.

RESULTS

Genetic diversity parameters

The 14 SSR loci amplified 146 different alleles, of which 141 were detected in the 10 wild populations (209 individuals), and 48 in the three crop populations (42 individuals) of brinjal. The number of private alleles for the wild brinjal pool was 107 and that of the cultivated counterpart was 7. Across the 14 loci, we detected a mean of 11.1 alleles per locus in the wild populations and 3.9 in the cultivars.

The number of alleles detected in wild populations, which were labeled as 1w–10w, ranged from 29 (9w) to 72 (6w). Mean allelic richness (Ar) in wild brinjal ranged from 2.07 (8w) to 4.65 (6w) and differed significantly among the populations (Kruskal–Wallis: ?2 = 22.7, df = 9, P = 0.007). Statistically significant differences were also found among the wild populations for the other genetic diversity parameters: mean private allelic richness (Ap; range: 0.04–0.60; P = 0.02), gene diversity (He; range: 0.199–0.554; P = 0.001) and observed heterozygosity (Ho; range 0.043–0.294; P = 0.011). For the crop populations, labeled as 1c–3c, the number of alleles varied from 26 in 2c (N = 10) to 45 in 1c (N = 24), whereas the mean allelic richness ranged from 1.68 in 2c to 2.28 in 1c. The trend was similar among the three crop populations for all the other genetic diversity parameters tested.

Wright's fixation index, FIS, a measure of inbreeding, was calculated for each population. Estimates of FIS for the 10 wild populations ranged from 0.505 (5w) to 0.905 (3w), with a mean FIS value of 0.752. Correspondingly, estimates of outcrossing rate (t) ranged from 5.0% (3w) to 32.9% (5w), with a mean t value of 14.6%. A Kruskal–Wallis test found significant differences in outcrossing among the wild populations (?2 = 19.9, df = 9, P = 0.018). For the three crop populations, 1c–3c, estimates of FIS were 0.627, 0.601, and 0.311, corresponding to estimated outcrossing rates 22.9%, 25.0%, and 52.5%, respectively.

Population structure and genetic relationships

Results of AMOVA showed significant genetic structure in the sampled populations, with 46% of the total variation explained by differences among populations and 54% by differences within the populations. Likewise, the global FST, a measure of population differentiation, was 0.48 (95% CI: 0.414–0.569; P < 0.001) among the 10 wild populations, intermediate between complete panmixis (0.00) and complete differentiation (1.00). Pairwise FSTvalues between populations ranged from 0.045 (5w vs. 6w) to 0.769 (4w vs. 8w), and all estimates were significantly greater than zero (P < 0.001). A test for isolation by distance further revealed a significant positive correlation (r2 = 0.15; P = 0.003) between genetic distance (FST/[1 - FST]; Rousset, 1997) and the natural logarithm of the geographic distance among the 10 wild populations. Bayesian-based STRUCTURE analysis and principal component analysis provide complementary methods for visualizing patterns of genetic similarity and differentiation among the 10 wild populations and three cultivated populations.

According to Evanno's ?K method for STRUCTURE, the highest peak was detected at K = 3, followed by peaks at K = 5 and K = 7. These three clusters are also seen in the first two dimensions of the PCA and hierarchical cluster analysis, although other dimensions of the data also explained considerable variation.

One of the three major clusters is made up of four wild populations (4w, 5w, 6w, and 7w) in the same general region across the states of Tamil Nadu and Karnataka. These four populations remained together at K = 4 but were split into two subclusters at K= 5, such that populations 5w and 6w were grouped together, as were the more distant populations 4w and 7w. The second major cluster consisted of three wild populations (8w, 9w, 10w) in southern Tamil Nadu and one wild population (1w) in Karnataka. This cluster stood out as an intact genetic group in STRUCTURE with successive increases from K = 2 to K = 5, pointing to close genetic similarity among the assigned populations. The third cluster was made up of two wild populations (2w and 3w) and the three cultivars. This group separated into two subgroups that reflected wild vs. cultivar origins. Although cultivar 1c was collected at the northernmost site in Karnataka, adjacent to the wild population 1w, it grouped very closely with 2c and 3c, which in turn showed a close affinity with 2w and 3w.

Overall, with the exception of population 1w collected in the north in Karnataka, the wild brinjal populations appeared to cluster according to their geographic origin, and their affinities were consistent across the three genetic analysis approaches. A few individuals of population 1w were assigned (jointly or exclusively) with the wild and cultivar brinjal at sites 2 and 3. Population 1w occupied an intermediate position between clusters II and III along the second principal component, which may reflect genetic admixture and/or mixed ancestry.

DISCUSSION

Genetic diversity

We found low to moderate levels of genetic diversity within the 10 wild populations and three crop populations of brinjal. Within some wild populations, genetic diversity was greater than that of the cultivars, and overall the wild populations had an average of 11.1 alleles per locus vs. 3.9 for cultivars. These results are consistent with two previous studies showing that wild brinjal harbored more genetic variation across its geographic range than cultivars. Reduced genetic diversity in cultivars is a common feature of many crops compared with their wild relatives, presumably due to genetic bottlenecks that occur during domestication and breeding.

Outcrossing rates

Variation in outcrossing rates may reflect differences in floral traits, availability of pollinators, population size and density, and other factors. We found significantly variable levels of natural outcrossing in the 10 study populations, based on estimates from FIS.

The three northernmost wild populations had the lowest outcrossing rates (5–8%). These populations were unusual in that 1w grouped with the southernmost populations, perhaps reflecting long-distance dispersal, while 2w and 3w were grouped with local cultivars, possibly due to a history of crop–wild gene flow. In contrast, all other populations had outcrossing rates of at least 13% and one, population 5w at Chadapatti, had over 400 plants and an outcrossing rate of 32%. A related study by our group also found a high degree of stigma exsertion in this population. Stigma exsertion was shown to influence outcrossing rates in natural populations of other solanaceous species such as S. ptycanthum (Hermanutz, 1991) and S. pimpinellifolium. In summary, wild brinjal has a mixed mating strategy in which selfing and outcrossing occur in various proportions across populations.

We were surprised that outcrossing rates of the cultivars were as high as 52.5% in population 3c (a home garden), and 23–25% in populations 1c and 2c (both small farms), respectively, given that many eggplant cultivars worldwide are capable of autogamous self-pollination. Estimates of outcrossing rates for cultivated brinjal based on some very early studies ranged from 2 to 46.8% (mean = 6.8%) in Japan (Kakizaki, 1924) and from 1.9 to 10.9% (mean = 6.9%) in India. Given that we sampled only three cultivated populations in this study, we are not able to draw general conclusions about outcrossing rates in the crop.

Patterns of population structure

Fifty-four percent of total genetic variation was found within wild brinjal populations, with 46% shared among populations, which is consistent with a mixed mating strategy and occasional long-distance gene flow. This level of differentiation among populations indicates strong divergence, and the degree of divergence was positively correlated with geographic distance over a range of ~600 km. This pattern conforms to the classical isolation by distance model of differentiation of Wright (1943), which predicts that the genetic distance between pairs of populations will increase as a function of the geographic distance between them due to limited gene dispersal by pollen or seeds.

Further, our analysis of population structure showed that the 10 wild brinjal populations could be separated into three major genetic groups, largely according to their geographic origin. This finding reiterates that geographic isolation and limited gene flow has played a significant role in shaping the contemporary patterns of genetic differentiation for wild brinjal in southern India. Nevertheless, in notable contrast, one wild population from Karnataka (1w) did not appear to cluster according to geographic provenance but showed genetic affinity with wild brinjal from Southern Tamil Nadu region, some 600 km away.

The similarity between population 1w and more southern populations suggests that long-distance founding events, possibly mediated by humans, may also have contributed to shaping the current genetic structure of wild brinjal populations.

Potential for crop-to-wild gene flow

Our analysis did not show evidence of distinct divergence between wild and cultivated brinjal. For example, based on the Bayesian STRUCTURE analysis, divergence between the two types of brinjal was not evident at two levels of hierarchical genetic structure (K = 2). At K = 3, we found that two wild populations clustered with two crop populations that grew near them and with a third crop population from farther west. The crop populations formed a fairly cohesive group, but were distinguished from wild populations only at K = 4 andK = 5.

These results are consistent with previous investigations based on allozyme and RAPD markers (Karihaloo and Gottlieb, 1995; Karihaloo et al., 1995), showing close genetic similarity between cultivars and related wild and/or weedy forms in India and supporting the view that cultivated brinjal and its wild and weedy relatives are conspecifics (Deb, 1989; Lester and Hasan, 1991; Karihaloo et al., 1995; Weese and Bohs, 2010). In contrast, Knapp et al. (2013) consider these taxa to be different species based on distinguishing characteristics of the stems, leaves, flowers, and fruits, which can be used to assign species to putative hybrids.

Our results also suggest that gene flow via pollen and/or seeds among sympatric crop–wild populations may have contributed to the observed genetic structure of brinjal in some cases (e.g., for populations 2w and 3w). Interestingly, these two populations had white immature fruits that are more typical of some cultivars, whereas other wild populations had immature fruits that were green or green with pale green patterns. White immature fruit color is a recessive trait (Daunay, 2008), and these populations had low outcrossing rates and high fixation indices.

Additional evidence suggesting shared ancestry between wild and cultivated brinjal, where a few individuals from wild populations other than

2w and 3w actually clustered with the crop and crop–wild groups, rather than the majority of wild plants from the same population.

Similar speculations about natural hybridization between cultivars and wild forms of brinjal have been made by previous authors (Omidiji, 1982; Karihaloo et al., 1995; Karihaloo and Rai, 1995). In addition, Davidar et al. (2015) report that wild and cultivated brinjal can co-occur, share pollinators, and produce viable F1 hybrid progeny, and M.-C. Daunay et al. (unpublished manuscript) show that F1 hybrid progeny exhibit heterosis. Therefore, hybridization and introgression of crop alleles into free-living populations may be ongoing in this crop–wild–weed complex, via pollen- and/or seed-mediated gene flow.

Implications for conservation of wild germplasm

Our analyses revealed a high degree of genetic divergence among populations of wild/weedy brinjal from southern India, given that about half of the total variation in the data were found among populations. The three major genotypic clusters that we detected in the sampled area can be used as a starting point for germplasm conservation.

Based on current knowledge, the most unique genotypic clusters are those labeled as cluster I and cluster II, while the two wild populations in cluster III had multilocus genotypes that were most similar to cultivars. The type of population-level sampling that we carried out is needed to inform strategies for the conservation and management of the surviving natural populations of wild and/or weedy relatives of cultivated eggplant. We expect that more extensive sampling of wild brinjal in other regions of India would reveal many additional target populations for germplasm conservation, as well as possible evidence of crop–wild gene flow.

In other studies of crop relatives, hybridization resulting from proximity to sexually compatible crops has led to changes in the diversity and/or distinctiveness of wild gene pools. Bartsch et al. (1999) and Song et al. (2006) reported greater diversity due to the acquisition of crop alleles in sympatric populations of wild beet and wild rice, respectively, while Kiang et al. (1979) attributed the extinction of remnant populations of wild rice in Taiwan to overwhelming gene flow from the crop.

Isolation from cultivars is generally considered to be beneficial from the standpoint of preserving the integrity and diversity of wild gene pools. The fact that two wild populations grouped very closely with cultivars suggests that there may be ongoing gene flow between them via hybridization and/or cultivars becoming naturalized as free-living populations. Also, Davidar et al. (2015) reported that 14 of 23 wild brinjal populations in southern India, including, occurred within 0.5 km of cultivated brinjal (this occurred at sites 1–4, 6, and 10). By examining variation at SSR loci in the present study, we found evidence for considerable differentiation among wild populations, but further sampling of local cultivated populations would be needed to evaluate which of these wild populations are sufficiently isolated from cultivated brinjal to maintain fairly distinct gene pools.

In conclusion, our study of the genetic structure of wild brinjal populations in southern India provides new evidence to support in situ and ex situ germplasm conservation strategies for this important crop relative. The Western Ghats area is recognized as an important in situ repository of germplasm for other cultivated species, particularly spices, and is considered to be the center of origin for the black pepper. Our findings support previous surveys suggesting that this region of southern India harbors a rich source of indigenous germplasm diversity for a wide range of crop plants.

CYTOPLASMIC-GENETIC MALE STERILITY GENE PROVIDES DIRECT EVIDENCE FOR SOME HYBRID RICE RECENTLY EVOLVING INTO WEEDY RICE

Weedy rice (Oryza sativa) has strong seed-shattering and seed dormancy and is highly competitive with cultivated rice (O. sativa). At present, there is no effective and selective chemical control available to cull the weedy rice population, given that both cultivated and weedy rice are the same species and share similar morphological and physiological characteristics. Weedy rice has therefore become one of the most harmful weeds in paddy fields. Based on the analysis of biological characteristics, molecular markers, and specific genes in plants from different areas worldwide, it has been proposed that weedy rice originated as hybrid progeny of wild rice (Oryza rufipogon) and cultivated rice, de-domestication or reversion of cultivated rice towards the wild type, or as hybrid progeny of indica-japonicasubspecies9. The origin and evolution pattern of weedy rice most likely differs among plants growing in different areas and should be explored.

Weedy rice has become a serious problem in China. Comparative studies of the morphological characteristics and genetic diversity of Chinese weedy rice indicate that it is closely related to companion cultivated rice and may indeed have originated by the reversion of cultivated rice or by gene introgression among rice varieties. We here provide new compelling evidence of the involvement of cultivated rice, in our case, hybrid rice, directly in the evolution of some weedy rice accessions.

Gene flow and introgression are driving forces for natural and cultivated plant evolution. They frequently and widely occur among cultivars, wild relatives, and compatible weedy types. These natural phenomena have existed for thousands of years and contribute to the evolutionary history of crops and their wild relatives. Many studies have reported that gene flow can occur among wild rice, cultivated rice and hybrid rice accessions. Existing evidence depicts strong relationships between cultivars and weedy rices at local levels to which our study contributes further proof that cultivated hybrid rice may directly participate in the evolution of weedy rice.

Rice has been cultivated in China for thousands of years but the spread of weedy rice is relatively recent. Coincidentally, hybrid rice rapidly became popular in China in the 1970?s, and by the 1990?s accounted for approximately half of the area of cultivated rice. Currently, some 17 million ha of hybrid rice are grown in China, representing 80% and 3% of its cultivated indica and japonica rice, respectively.

Unlike the other two forms of rice widely cultivated in China - landraces (local varieties selected and maintained by farmers) and conventional varieties (those improved by breeders) - the continued production of which depends on their autogamy, hybrid rice is instead the heterotic F1generation of crosses between genetically distant parents. Thus, characteristics of hybrid rice progeny often strongly segregate owing to the great genetic diversity of parental lines. Hybrid rice progeny that escapes harvesting may segregate into distinct biotypes through gene recombination, some of which might evolve into weedy rice.

Interestingly, Jiangsu province is currently experiencing a serious infestation of weedy rice, most of which is of the indica-type despite the fact that japonica conventional varieties have been the primary type of cultivated rice in this region for more than 20 years. However,indica-type hybrid rice was planted on a large scale in this province in past decades. It is suspected that weedy rice in some areas of Jiangsu province may have evolved from segregating progeny of indica hybrid rice, which is supported by results from genetic and morphological analyses. Despite the widespread planting of hybrid rice, little is known about the contributions of its segregating volunteers to the ongoing evolution of weedy rice. Resolving the evolutionary origin of weedy rice in areas of hybrid rice cropping is particularly relevant, as any involvement of hybrid rice in the spread of weedy rice would pose challenges to the sustainability and efficiency of hybrid rice technology.

Hybrid rice is obtained by either three-line or two-line systems. The former was developed earlier and became widely grown, achieving more than 90% of the hectarage dedicated to hybrid rice between 1976 and 2000. Since then, its proportion has decreased only slightly to approximately 80%. The female parent of three-line hybrid rice is a cytoplasmic male sterile line, whereas the male parent is a restorer line. The hybrid contains cytoplasmic male sterility (CMS) genes and the heterozygous restorer of fertility (Rf) genes.

The male-sterile female parent can be traced by cytoplasmic genotype analysis due to maternal inheritance of the sterility gene. Although the potential for a minor paternal transmission of the mitochondrial genome has been recognized in several species and the ultrastructure of the rice sperm cell shows mitochondria localized mainly in the head of the sperm cell, organelle DNA was not detected. Thus, it is reasonable to assume that the CMS gene is primarily maternally inherited. If weedy rice directly evolved as progeny of hybrid rice, weedy rice individuals should contain a CMS gene regardless of any involvement of gene flow during the evolution process. This offers a unique way to verify the direct involvement of hybrid rice in the origin of some weedy rice types.

A wild abortive (WA) cytoplasmic male sterile line was first developed and became the most widely used to breed indica hybrid rice. A Boro type (BT) cytoplasmic male sterile line is the most widely used to breedjaponica hybrid rice. To test the hypothesis that some weedy rice accessions may have directly originated from hybrid rice, we used the CMS-WA and CMS-BT genes as markers to test the lineage of a collection of weedy rice samples. These genes are largely maternally inherited and are therefore not expected to be transferred to weedy rice through pollen-mediated gene flow.

RESULTS

Indica-japonica subspecies classification and cross-compatibility of weedy rice accessions

Prior to assaying 322 weedy rice accessions collected across China for CMS genes, 36 accessions independently obtained from three major rice-producing provinces experiencing serious weedy rice infestations (Liaoning, Jiangsu and Guangdong) were assigned to indica or japonicasubspecies according to Cheng's Index, which measures six rice characteristics. This allowed us to rule out infertility associated withindica-japonica hybridization. Of the 12 weedy rice samples from Liaoning province, eight accessions were japonica and four werejaponica-cline types. Conversely, of the 24 weedy rice accessions from Jiangsu and Guangdong, eight accessions were indica and 16 wereindica-cline type. Thus, there was regional differentiation inindica-japonica subspecies. Two-thirds of the weedy rice samples (i.e., those identified as japonica- and indica-cline types) can be considered intermediate forms between indica and japonica rices.

Sixty-nine hybrid progeny were obtained from 72 artificial crosses between the 36 weedy rice (WR) accessions and either the indica-type WA and japonica-type BT maintainer lines LTPB and S28B, respectively, as pollen donors, (36 WR × LTPB, S28B). No hybrids were obtained from the combinations of three accessions from Guangdong (GD) with the S28B line (GD1,GD3,GD4 × S28B) due to seed abortion before grain filling. All the obtained hybrid progeny were fertile according to affinity detection tests except for GD9 × S28B, which had very low fertility. Therefore, most combinations (all weedy rice × LTPB and most weedy rice × S28B) were cross-compatible, with the exception of the combinations of four of the GD weedy rices × S28B. The cross-incompatibility observed in these four combinations indicates that GD1, GD3, GD4 and GD9 belong to the indica subspecies. Weedy rice from Liaoning (LN) hybridized better with S28B than with LTPB, reinforcing the notion that weedy rice from this region is closely related tojaponica. Conversely, weedy rice accessions from Jiangsu (JS) and Guangdong were apparently more closely related to indica because hybridization was more effective with LTPB than with S28B. Most weedy rice accessions, however, were widely compatible because they were cross-compatible with both typical japonica and indica rice, and the difference between the affinities was not considerable. The classification of weedy accessions based on hybridization affinity was consistent with the classification obtained with Cheng's indices.

The relationship between weedy rice and BT hybrid rice

The seed setting rates of the hybrid progeny of weedy rice accessions and the sterile line S28A as the female parent (S28A × 36WR) were all less than 5%, which suggested that these weedy rice accessions lack the Rf gene for CMS-BT. It is unlikely that they also have the CMS-BT because they set seed normally. As artificial (GD1, GD2, GD3, GD9 × S28B) hybridizations were cross-incompatible, and given that maintainer line S28B shares the same nuclear constitution as S28A, the infertility of these hybrid progeny might involve other factors in addition to the absence of the Rf genes for CMS-BT in the weedy rice parents.

Assay of pollen viability in hybrid progeny (S28A × 36WR) by I2-KI solution was not sufficiently clear to ensure accuracy, because the pollen of BT type male sterile lines typically has spherical abortion characteristics or a stained abortion phenotype. The size and staining depth of pollen grains were diversiform.

The reliability of the p3p4 marker, specific for CMS-BT, was verified and it was determined that it only amplified with the DNA fragment from material containing the CMS-BT gene. None of the 36 weedy rice accessions contained the CMS-BT gene, according to assay of the p3p4 marker. These results are consistent and suggest that these weedy rice accessions are not genetically related to BT hybrid rice.

RELATIONSHIP BETWEEN WEEDY RICE AND WA TYPE HYBRID RICE

Nine hybrid combinations between the male-sterile line LTPA as female parent and weedy rices JS3 and JS10 (Jiangsu) and GD2, GD4, GD7, GD9, GD10, GD11, and GD12 (Guangdong) as male parents were fertile as determined by pollen viability and seed setting of their hybrid progenies. The infertility of the remaining hybrid progenies must have been under the control of a CMS-WA gene (weedy rice does not contain the Rf genes for CMS-WA). No false negatives existed; this was concluded because LPTB shares the same nuclear constitution with LTPA, and there was no cross-incompatibility in the artificial 36WRxLTPB hybridizations. Therefore, the nine weedy rice parents must have contained the Rf gene for CMS-WA (abbreviated to R-WA) to produce fertile hybrids with LTPA. Thus, they were subjected to further analyses to determine whether they also carried the CMS-WA.

Four combinations of the nine F2 hybrids (9 R-WA × LTPB) of interest, segregated sterile plants whose female parents were GD9, GD10, GD11, GD12. The F2 hybrid progeny with GD11 as the female parent conformed to a 1:3 ratio of sterile plants to fertile plants according to a chi-square test for segregation ratio, and those for which the female parents were GD9, GD10, and GD12 exhibited a 1:15 ratio. Different segregation ratios could be attributed to the weedy rice accessions containing one or two main effect Rf genes for CMS-WA. This result confirmed that weedy rice accessions GD9, GD10, GD11, and GD12 contained CMS-WA and had a strict direct relation to WA type hybrid rice. Coincidentally, the cultivated rice accessions grown together with these four weedy rice lines were three-line hybrids containing CMS-WA.

IDENTIFICATION OF WILD ABORTIVE CYTOPLASMIC MALE STERILITY IN OTHER WEEDY RICE ACCESSIONS

Putative cms marker (specific marker to detect CMS-WA) was considered validated if it reliably yielded a band only for the CMS-WA gene that was contained by the plant. According to an assay for the cms marker, 12 of the 322 additional weedy rice accessions contained CMS-WA. One of these 12 accessions was from Jiangsu province, two accessions were from one population from Hainan province, and nine accessions were from three populations from Guangdong province. These results were further verified by artificial hybridization as described above. All progeny of the 12 F2hybrids tested (12WR containing CMS-WA detected with the cmsmarker × LTPB) segregated sterile plants. These results are consistent and confirm that these accessions contained CMS-WA and were genetically related to a WA-type hybrid rice.

CORROBORATION OF RESULTS

During the process of preparing this manuscript, the gene WA352 that is responsible for CMS-WA was cloned. Using specific marker orWA352 based on WA352 to detect the possible presence of CMS-WA in all 358 weedy rice samples, we unequivocally corroborated our previous findings with the cms marker: the same 16 of the 358 weedy rice samples were confirmed as having

CMS-WA. Furthermore, we sequenced the 16 amplified products and consistently found them containing the WA352 gene.

DISCUSSION

SUBSPECIES AFFILIATION OF SELECTED WEEDY RICE ACCESSIONS AND THEIR RELATIONSHIP TO HYBRID RICE

Individuals from a selected collection of 36 weedy rice accessions - 12 from each of three Chinese provinces - were assigned to indica andjaponica subspecies and used to validate experimental methods. All weedy rice accessions from Liaoning, where japonica rice is predominantly grown, belonged to this subspecies (eight accessions) or were identified as japonica-cline (four accessions) but none were derived from hybrid rice. Previous studies have determined that weedy rice accessions from this northern province are genetically narrowly diverse (within populations) and are closely related to cultivatedjaponica rice but distant to indica rice. They probably originate from local varieties through de-domestication and hybridization. Our results are consistent with previous studies despite the limited sample size, and they rule out the involvement of hybrid rice in the origin of this province's weedy rice.

The weedy rice accessions from the other two provinces, Jiangsu and Guandong, were all indica (although predominantly indica-cline), also in agreement with previous studies. Currently, rice production in Jiangsu is primarily of japonicacultivars, although indica rice has been grown in the past and is still planted adjacent to japonica fields by some farmers. These conditions may help explain the intermediate nature of a portion of the tested weedy rices, which may have evolved through gene flow and hybridization. In Guangdong, where weedy rice is a serious problem exacerbated by increasing adoption of direct seeding, cultivated rice is primarily of the indica type. Weedy rice most likely originated from local cultivated rice and not from the japonica-type wild rice found in the province. Moreover, four weedy rice accessions from Guangdong - GD9, GD10, GD11, and GD12 - contained CMS-WA and were considered direct descendants of WA type hybrid rice.

HYBRID RICE HAS PARTICIPATED IN THE EVOLUTION OF SOME WEEDY RICE IN CHINA

Sixteen weedy rices (one from Jiangsu, two from Hainan, and nine from Guangdong), representing 4.5% of the total 358 accessions analyzed, contained CMS-WA. The accuracy of this finding is supported by both genetic and molecular marker tests. The CMS-WA that was originally selected from a sterile mutant wild rice plant (O. rufipogon) has been exploited in the production of the majority of three-line indica-type hybrid rice cultivars in

China. Hybrid rice containing CMS-WA began occupying increasing proportions of the area dedicated to cultivating rice in the 1970?s. The sterile phenotype in wild rice is extremely rare because the CMS gene, although probably present in a relative high proportion (up to 5%) of the wild individuals, is overcome by the restoration of fertility by Rf; the finding of the original sterile individual was indeed a major breakthrough in rice breeding. For wild rice to be a source of the CMS-WA gene the individuals carrying it should have participated as female parents in crosses with cultivated forms to generate weedy types.

A recently study has found a lack of relationship of weedy rice with wild rice in areas where O. rufipogon is still present. This suggests that the CMS-WA in weedy rice most likely was acquired recently (within the last 40 years) from hybrid rice. Because this gene is coded by the mitochondrial genome, it is primarily maternally inherited, indicating that the original hybrid parent was likely female. This provided direct evidence that hybrid rice has been involved in the evolution of some weedy rice accessions.

The majority of the 358 accessions came from the three provinces (Guangdong, Hainan, and Jiangsu) where weedy rice infestations are most severe. However, given the considerable number of accessions tested and their disparate geographical origins, it appears that weedy rices directly related to hybrid rice are a relatively small proportion of those currently infesting rice fields in China. There is a chance that the CMS trait is unfit for weedy rice and segregates out of weedy populations, further reducing the frequency of weedy rice individuals carrying the CMS gene. However, the CMS-WA phenotype can be restored by either of two non-allelic, dominant Rf genes. Thus for hybrid rice individuals containing one or two heterozygous Rf genes, the expected proportion of sterile individuals among its volunteers (the F2 generation) would theoretically fall between 0.25 (3:1) and 0.062 (15:1).

If the two Rf genes become homozygous or if there is a sufficient load of pollen containing Rf genes from cultivated or weedy rice (frequent in regions of hybrid rice cultivation), plants carrying the CMS-WA will become fertile and consequently the unfitness conferred by CMS-WA would be relieved. This would rule out the notion that the outbreak of weedy rice in China can be attributed to the ample cultivation of hybrid rice. Although CMS-BT has been exploited injaponica hybrid rice production, it was not detected in any of the 36 weedy rice accessions from Liaoning, Jiangsu and Guangdong. The area dedicated to cultivating BT-type hybrid rice is limited, thus decreasing the probability of detecting weedy rice accessions containing the CMS-BT.

Possible evolutionary processes leading to weedy rice

The evolution of some weedy rice accessions from cultivated hybrid rice has probably been a complex process. The CMS-WA in weedy rice could have come from hybrid rice through direct or indirect evolution. Direct evolution entails the possibility that some individuals of segregant populations of hybrid rice becoming weedy rice through gene recombination.

The heterosis in hybrid rice can be considered only transitory in the scheme of weedy rice evolution, but gene recombination in progeny should have a long-standing effect on adaptation to novel environments by the creation of new gene combinations upon which selection can act. A case study has shown that hybrid progeny of different rice varieties could evolve into weedy rice like individuals, and their occurrence was more frequent when the parent accessions were more genetically distant. Parents with rich genetic diversity are ordinarily selected for hybrid rice breeding to obtain the strongest heterosis. By the end of June 2014, 2918 three-line hybrid rice cultivars had been released in China. Such an amount of hybrid rice variability would produce numerous types of recombinants whose evolutionary direction would be unpredictable.

Establishment of hybrid rice progeny in the field depends mostly on volunteers. Farmers do not save the seeds of hybrid rice for future planting, as the benefits of the heterosis in the F1 hybrid are lost upon segregation in the F2 generation. However, an estimated 300?kg ha-1 of rice seeds can reach the soil after harvesting due to natural shattering and losses during harvesting. These seeds have the potential to establish as volunteers.

Their success in subsequent rice plantings (in continuous rice production systems) or in rotational crops is closely related to climate conditions and crop production systems. Weedy rices with a filial relation to WA type hybrid rice were mostly detected in Hainan and Guangdong. The conditions of suitable temperatures and continuous rice cultivation in these southern provinces of China are conducive to weedy rice evolution and persistent generation of volunteers.

Indirect evolution of weedy rice would be mediated by gene flow between hybrid rice or its progeny and other rice varieties, wild relatives or already existing weedy types. The outcome of this process would be determined by the outcrossing frequency, the direction of the gene flow and the genetic background of the accompanying rice forms in the paddy field itself or in adjacent fields within the range of gene flow. A cultivar could serve as a bridge for the movement of the CMS gene from hybrid rice to weedy rice; recently it has been determined that some indica cultivars contain the CMS gene.

Outcrossing rates can significantly affect the heterozygosity of a population. The genes related to environmental adaptation and competitive ability present in different rice types would be selected for over time once initial hybridization provides opportunity for recombination. Some genes for enhanced traits from rice, including transgenes, can introgress into weedy rice. Likewise, undesirable weediness genes from weedy rice can be transferred into cultivated rice. Independent studies have found that the F1, F2, and F3 progeny of hybrids between weedy rice and cultivated rice were at least as fit as the weedy rice parents, indicating that the hybrids have the potential to evolve into new types of weedy rice. Thus, if weedy rice is present within or nearby hybrid rice fields, new weedy rice types may arise more rapidly through pollen exchange with existing hybrid rice.

Gene flow between cultivated rice and weedy rice can occur in both directions, with an outcrossing rate usually below 1%. However, outcrossing may be affected by cultivar and weedy rice biotype. Hybrid rice may have a higher outcrossing rate due to its exserted stigma. Indeed the herbicide-resistant hybrid cultivar CLXL8 was a higher outcrosser than the conventional cultivar CL161.

Furthermore, hybrid rice for which a gametophyte abortion type sterile line is used, as in the case of CMS-BT, produces only approximately 50% viable pollen grains, which could affect its outcrossing rate. The outcrossing rate is also determined by factors such as plant distance, plant height, wind direction, synchronization of flowering and pollen viability. Hybrid rice volunteers segregate plants of diverse heights with extended flowering periods. In addition, for hybrid rice that uses a sporophyte abortion type sterile line, as in CMS-WA, there will be male sterile individuals among the progeny that would have to be cross-pollinated by fertile individuals to create offspring.

There is also a risk of hybrid rice seed contamination with weedy rice seed, because the majority of hybrid rice seed is produced in open farm fields and not at experimental farms. A significant proportion of plants in a seed production field are a sterile line that must be cross-pollinated. Thus, without proper isolation or in the presence of weedy rice arising from dormant seed, there is opportunity for gene flow to occur with the consequence of off-types (weedy rice) being planted along when the obtained hybrid rice seed is sown. This would accelerate the onset of the indirect evolution process previously discussed.

Inferences on the time of origin of weedy rice

The proportion of weedy rice accessions found to have a direct relationship to hybrid rice by this study is much lower than expected, considering the high potential of hybrid rice to evolve into weedy rice. This propensity is likely because hybrid rice has heterozygous segregating progeny (volunteers) with a potentially higher outcrossing rate. Additionally, comparative studies of weedy and cultivated rice in China have shown that either weedy rice has a close relationship with companion cultivated rice or that the crop was involved in weedy rice evolution through gene introgression. Thus, most of the current weedy rice accessions in China are related to conventional cultivated rice (including both landraces and conventional varieties). This poses the question of whether the current weedy rice biotypes evolved along with the introduction of hybrid rice or much earlier. Considering that hybrid rice and conventional cultivated rice have shared almost the same proportion of the rice cultivation area in China for several decades, one could surmise that most weedy rice accessions evolved from conventional cultivated rice much earlier, i.e., over thousands of years of rice cultivation. The worsening of the infestations currently observed is most probably related to agronomic factors, such as increased direct seeding, planting of contaminated seed and ineffective control measures. The regulations for rice seed certification in China address quality aspects including purity and germination, but weedy rice seed is not currently prohibited.

Furthermore, rice production in China has long depended on abundant landraces selected and kept by farmers since crop domestication began. In recent decades, landraces have been gradually replaced by improved varieties. During the Seventh Five-Year Plan period (1986-1990) of China, 34663 rice landraces were collected and stored, of which 8963 - 26% of the collection - had a red pericarp. Several red pericarp landraces were tall, matured early and were tolerant of poor soils and drought. Historical records from Jiangsu also describe local rice cultivars as having long awns, a dark husk, a red pericarp and other traits frequently associated with weedy rice. It is thus possible that landraces were an important source of the Rc (red pericarp) gene and of other weedy traits involved in weedy rice evolution.

Weedy rice continuously evolves with the cultivated rice it infests. Since hybrid rice cultivation is very recent, weedy rice evolving from it should be at an early stage of evolution. It is expected that with increased planting of hybrid rice, opportunities for weedy rice to directly or indirectly evolve from hybrid rice and containing the CMS gene will also increase. Long-term dynamics of weedy rice evolution from hybrid rice would also be modulated by outcrossing rates and by the contribution of Rf genes that are not always homozygous. Special attention should be paid if in the future, transgenic hybrid rice, particularly carrying herbicide resistance traits, is released, as its progeny might also have high outcrossing rates, and the new traits will alter the dynamics of weedy rice evolution under the selective pressure of the herbicide.

METHODS

WEEDY AND CULTIVATED RICE MATERIALS

Thirty-six weedy rice accessions initially collected (12 each) in Liaoning, Jiangsu and Guangdong provinces were selected from the weedy rice germplasm bank of the Weeds Research Laboratory at Nanjing Agricultural University (NJAU) and used in field experiments for CMS detection. An additional 322 weedy rice accessions from 91 populations from 12 provinces where hybrid rice was or is widely planted were included in the molecular tests. All weedy rice samples were collected from populations infesting commercial rice fields by selecting a single panicle per plant of at least 20 plants separated by a distance of at least 10?m in each location. We randomly selected 1-5 individuals from these samples, representing each population. Eight pairs of sterile and maintainer rice lines (provided by the Rice Institute of NJAU) were used as parents for artificial hybridization and as controls for CMS markers.

General experimental approach

Our experimental approach to determine whether hybrid rice is implicated as a direct source of some of the weedy rice biotypes currently infesting rice fields in China.

It involved the detection in weedy rice of the most important CMS genes used in the production of hybrid rice cultivars as a diagnostic trait for their direct parenthood of weedy individuals. The japonica and indica affinity of the 36 weedy rice accessions was determined through morphological characterization and analysis of artificial crosses between the weedy forms and appropriate maintainer lines for pollen viability and seed setting. Artificial hybrids with sterile lines were also used to determine the presence of CMS and Rf genes in the weedy types.

Artificial hybridization and fertility detection

The warm water emasculation method was used for artificial hybridization. Selected panicles about to bloom were submerged for five minutes in warm water at 43?°C and 45?°C for indica and japonicarice, respectively, to kill the stamens without injuring the pistils. One-third to a half of each glume was carefully cut to expose the pistils. A paper pollination bag with both ends open was placed over each prepared panicle and closed at the bottom with a clip. While maintaining the opening at the upper end, pollen of the designated donor was sprinkled on the pistils, and the bag was closed.

The fertility of experimental materials was assessed by a pollen viability test and by determining the rate of seed set. Mature florets, with filaments about half the length of the glumes, were selected to detect pollen viability by I2-KI staining, either immediately or after being temporarily preserved in formaldehyde acetic acid solution.

The stained samples were mounted on glass slides and observed by optical microscopy (Olympus Co., Tokyo, Japan). Three florets from each individual plant were selected. Each floret was observed in five vision fields containing approximately 100 pollen grains each. The ratio of stained to unstained pollen grains was determined. A healthy panicle was selected from each plant and self-crossed by enclosure in a porous non-woven fabric bag before flowering for determining seed set at maturity.

Validation of a molecular marker of the male sterile gene

Fresh leaves were taken from each individual plant for total DNA extraction using plant genomic DNA rapid extraction kit (HF213-01, Yuanpinghao Biotech Co., Ltd. Tianjin, China). CMS-BT marker p3p4 (p3: 5'-ATGGCAAATCTGGTCCGATG-3' and p4: 5'-ACT TAC TTA GG AA AG AC TAC-3') and CMS-WA marker cms (cmsF: 5'-ACTTTTTGTTTTGTGTAGG-3' and cmsR: 5'

TGCCATATGTCGCTTAGACTTTAC-3') were used. Polymerase chain reaction (PCR) was performed in an automated PCR apparatus (BioRAD PTC300) using a 25?μL amplification reaction system as follows: 1.0?μL DNA template at 10?ng μL-1, 12.5?μL of Premix Taq containing 0.625 U of DNA polymerase, 0.4?mM of each dNTPs and 2 × Taq buffer (TaKaRa Biotechnology Co., Dalian, China), 1.0?μL each of the forward and reverse primers at 10?μM and 9.5?μL of distilled water. The PCR procedure included an initial denaturation for 5?min at 94?°C; 30 cycles of 94?°C for 1?min, 55?°C for 1?min and 72?°C for 1?min; extension for 10?min at 72?°C and maintenance at 4?°C.

Subspecies classification and affinity determination

Classification of weedy rice accessions into indica and japonicasubspecies is important to better understand their genetic relationships with the accompanying cultivated rice. Subspecies identification was accomplished using the Cheng's index protocol.

Sterile hybrids in crosses between sterile lines and weedy rice can occur because of indica-japonica cross-incompatibility or lack of the Rf genes in weedy rice. Thus, it was important to classify the accessions into the two subspecies and to test the affinity with typical indica(LTPB) and japonica (S28B) rices before detecting the Rf genes by artificial hybridization. The 36 weedy rice accessions were selected as female parents, and indica type WA maintainer line LTPB and japonicaBT type maintainer line S28B were selected as pollen donors (72 hybrid combinations). Ten plants of each combination were used to determine pollen viability and seed setting rate. The differences in fertility of F1hybrids (36WR × typical indica, 36WR × japonica rice) were used to determine the incompatibility between weedy rice accessions and sterile lines. These were used in the detection of the Rf genes to exclude false negatives and also to categorize the subspecies of weedy rice as indica- or japonica-like. Seed produced by selfing the F1 hybrids was used to detect CMS.

Detection of cytoplasmic male sterility and restorer of fertility genes

Artificial hybridization is a reliable way to detect the presence of Rf and CMS genes in weedy rice accessions. Plants of the 36 weedy rice accessions were selected as male parents; WA type sterile line LTPA and BT sterile line S28A were used as female parents (resulting in 72 hybrid combinations). Pollen viability and seed setting rate of these F1hybrids were determined. Weedy rice individuals without the Rf genes (those whose offspring pollen aborted and did not set seed) were eliminated because it was highly unlikely for them to carry the CMS-BT or CMS-WA.

The CMS-BT is gametophytic, and the F2 hybrid generation of a BT type male sterile line and a restorer line will only produce fertile offspring. Gene orf79 is responsible for CMS-BT. Marker p3p4 can amplify a 239?bp fragment which is part of orf79 from the BT sterile line and its hybrid progeny, but cannot amplify a fragment in a maintainer line because it lacks CMS-BT. Four pairs of BT sterile and maintainer lines were used to verify the reliability of this marker before using it to detect the CMS-BT in weedy rice.

The fertility of F2 progeny of which weedy rice was the female parent (containing the Rf genes for CMS-WA) and LTPB was male parent was determined. At least two hundred plants from each hybridized combination were randomly selected for tests of pollen viability and seed setting rate. Those plants identified as sterile were considered to be derived from the progeny of weedy rice accessions used as female parents containing the CMS-WA.

The molecular marker cms can amplify a 386?bp fragment from the WA sterile line and the derived hybrid, but not from the maintainer line. It was selected for detecting the CMS-WA in the additional 322 weedy rice samples. The reliability of the cms marker was determined with WA sterile lines, WA maintainer lines and the weedy rice accessions confirmed to contain the CMS-WA. Artificial hybridization was also used to further guarantee the accuracy of detection with the molecular marker. Weedy rice accessions bearing the cms molecular marker for CMS-WA were used as female parents in crosses with LTPB as pollen parents. Fertility determination in the segregating F2 generation (at least two hundred plants from each hybridized combination) was used to assay whether the CMS-WA existed in the weedy rice parents.

Corroboration of presence of cytoplasmic male sterility by specific molecular markers

In the course of analyzing the data and preparing the manuscript,WA352 the gene responsible for CMS-WA was cloned31. In order to confirm the reliability of our procedures and the accuracy of our results, we used marker orWA352 (orWA352-F: 5'-GTTGATGGGTATGGATAGAG-3' and orWA352-R: 5'-CGCAGGGCCTCGGTATATCTA-3') based on WA352 to detect the possible presence of CMS-WA in all 358 weedy rice samples.

Bibliography

A V S S Sambamurty: *Taxonomy of Angiosperms*, I K International Publication, Delhi, 2005.

A.V.S.S. Sambamurty and N.S. Subrahmanyam: *Economic Botany of Crop Plants*, Asiatech Publication, New Delhi, 2000.

Agnes Arber: *Water Plants : A Study of Aquatic Angiosperms*, BSMPS Books, Delhi, 2003.

B.P.Pandey: *A Text Book of Botany : Angiosperms*, S. Chand Publisher, Delhi, 2006.

Balwant Kumar; Sudhir Chandra; Kiran Bargali and Y P S Pangtey: *Ethnobotany of Religious Practices in Kumaun (Havan)*, Bishen Singh Mahendra Pal Singh, Delhi, 2007.

Chahal, S.S.; I.B. Parashar: *Achievements and Prospects in Mycology and Plant Pathology*, International, Delhi, 1997.

Chaturvedi, G.S. : *Abiotic Stresses and Plant Productivity*, Aavishkar Pub, Delhi, 2010.

Daubenmire, R. F.: *Plants and Environment*: New York, Wiley, 1947.

Grabley, S. and R. Thiericke : *Drug Discovery from Nature*. Berlin, New York, Springer, 1999.

Jackson, E. : *Crop Management and Soil Conservation*, Biotech Books, Delhi, 2011.

Kannaiyan, S. : *Bioresources Technology for Sustainable Agriculture*, Associated Pub., Delhi, 1999.

Keshab R. Rajbhandari: *Ethnobotany of Nepal*, EthnoBotanical Sciences/anical Soc of Nepal, 2001.

Kumar, U. : *Methods in Plant Tissue Culture*, Agrobios, Delhi, 2011.

M V Patil and D A Patil: *Ethnobotany of Nasik District, Maharashtra*, Daya Publication, Delhi, 2006.

N.C. Mukhopadhyay: *Plant Taxonomy*, Aavishkar Publication, Delhi, 2006.

N.P. Singh, K.K. Khanna, V. Mudgal and R.D. Dixit: *Flora of Madhya Pradesh: Vol. III. Angiosperms (Hydrocharitaceae to Poaceae) and Gymnosperms*, Botanical Survey of India, 2001.

P P Sharma & N P Singh: *Ethnobotany of Dadra, Nagar Haveli and Daman (Union Territories)*, Botanical Survey of India, 2001.

Pooja: *Economic Botany*, Discovery Publishing House, Delhi, 2005.

R.A. Raju: *Ethnobotany of Rice Weeds in South Asia*, Today Tomorrow's Publication, Delhi, 1999.

R.K. Maiti and V.P. Singh: *An Introduction to Modern Economic Botany*, Agrobios Publication, Jaipur, 2006.

Rajendra D Kshirsagar and N P Singh: *Ethnobotany of Mysore and Coorg, Karnataka State*, Bishen Singh Mahendra Pal Singh Publication, Delhi, 2007.

Ramesh Umrani: *Basics of Economic Botany*, Anmol Publication, Delhi, 2009.

Rashtra Vardhana: *Economic Botany*, Sarup Book, Delhi, 2009.

S.K. Sood and Smriti Thakur: *Ethnobotany of Rewalsar Himalaya*, Deep and Deep Publication, Delhi, 2004.

S.K. Varma, D.K. Sriwastawa and A.K. Pandey: *Ethnobotany of Santhal Pargana*, Narendra Publication, Delhi, 1999.

Shubhangi Pawar and D A Patil: *Ethnobotany of Jalgaon District, Maharashtra*, Daya Publication, Delhi, 2008.

Index

A

Antioxidant Stability 57
Arsenic Concerns 25

B

Bacillus Cereus 25

C

Calcium Absorption 57
Climate Matching 113, 122
Companion Plant 29
Cropping Systems 168, 178
Cultivated Rice 178, 179, 180, 181, 182, 183, 184, 187, 188, 189, 190, 191, 192, 193, 194, 195, 196, 197, 201, 203, 206, 207, 210, 211, 212, 215, 216, 217, 218, 219, 239, 240, 243, 244, 247, 248, 250

D

Drug Development 2, 4

E

Economic Botany 12, 131
Entrenchment Phase 100
Environmental Impacts 38
Equivalent Method 85
Establishment Phase 99
Ethnobotany 1, 2, 3, 12, 13, 14, 15, 16

F

Food Plants 12, 16, 117, 118, 119, 120, 121, 123, 124, 125, 126, 127, 128, 129, 130, 131
Future Potential 44

G

Golden Rice 44, 47, 48, 49, 50, 51, 52

H

Health Benefits 56, 58
Hybrid Rice 37, 228, 239, 240, 241, 242, 243, 244, 245, 246, 247, 248, 249

I

Inverted Bottle 74, 75

M

Management Practices 127, 131, 139, 142, 145, 170, 171
Management Variables 139, 140

P

Pest Management 40, 41, 77, 171
Phenotypic Selection 219, 220, 223, 226, 227
Plant Material 14, 95, 174, 198
Puffed Rice 52, 53, 54, 55

R

Regional History 26
Research Population 119

S

Seed Identification 69, 70, 71
Seed Packet 74, 75
Statistical Analysis 175, 186, 192, 226
Subsequent Development 49

T

Transplanting Techniques 170

W

Weed Biodiversity 117, 137, 145, 146
Weed Communities 63, 140, 143
Weed Evolution 196, 208, 210
Weedy Rice 178, 179, 180, 181, 182, 183, 184, 187, 188, 189, 190, 191, 192, 193, 194, 195, 196, 197, 198, 199, 200, 201, 202, 203, 204, 205, 206, 207, 208, 209, 210, 212, 215, 216, 217, 218, 219, 220, 221, 227, 228, 239, 248, 249, 250, 251
Wild Germplasm 197, 231, 238
Worldwide Consumption 38